100

LIVESTREAMING & DIGITAL MEDIA PREDICTIONS

VOLUME 3

Top Content Creators Help You Succeed Online in an Era of Rapid Change

ROSS BRAND

livestreamuniverse.com

First published by Livestream Universe
ROSS BRAND
100 LIVESTREAMING & DIGITAL MEDIA PREDICTIONS, Volume 3
Top Content Creators Help You Succeed Online in an Era of Rapid Change

First Edition 2023

THE 100 PREDICTIONS SERIES

#1 Amazon Best Seller in Six Countries

Winner of 18 Book Awards

JOIN OUR UNIVERSE

As a valued *100 Predictions* reader, you are invited – and encouraged – to become a part of our Livestream Universe community. We are planning exciting new initiatives to help you build your influence and monetize your content.

The best way to stay in the loop is to join our free email list:
livestreamuniverse.com/streamleader

The best way to connect with the book's author and contributors is to join our free Facebook Group:

facebook.com/groups/monetizelive

For my parents, Marion & Irv Brand

REACTIONS TO VOLUME 2

If you want to learn how to get far more visible and make a much better impact through digital media and livestreaming, these books are the only resource you need. Your Predictions books are AMAZING, Ross! You've truly outdone yourself with Volume 2... and I hope this signals that this will be an ongoing series. I love how this book provides a fun, super quick, easy to read, and most importantly, ACTIONABLE collection of predictions from a wide variety of livestreamers and digital media geniuses. I like to think of it as part cheat sheet to this online world and part rolodex of digital badassery. Each contributor offers a unique perspective for you to consider as you press on with your own digital media endeavors. It's the Who's Who of our space and time, and I am truly honored that Ross keeps inviting my perspective back to be included among such superstars - with Ross himself chief among us!

-Jennie Mustafa-Julock aka Coach Jennie

"Find out exactly what you need to know to create powerful and persuasive content for social media, web3 and the metaverse. Ross Brand and his contributing experts knock this one out of the park!"

-Mitch Jackson, 2013 California Litigation Lawyer of the Year & Co-Founder ManeuVR

"It is so helpful to understand the trends and reap the benefits of the wisdom of those who have been successfully creating top content for many years. Some of the highlights thoughts and predictions for me: Rob Greenlee, Judi Fox, Dave Jackson, Rocky Nash. So many gems. I can't wait for my hard copy to arrive. And I like how at the end of the book there are lists of classes to take, and ideal audio gear for whatever your niche is. It really helps take the guesswork out of new ventures and gives great ideas for wishlists."

-Meg Glesener, host of Letters from Home Podcast

"Ross Brand put together a great compilation of some of the best ideas from some really smart people. This is a book you will want to read and implement as soon as possible for your business. You will be glad you did it."

-Terry Brock, Hall of Fame Professional Speaker

"When you gather experts to give their advice on one of the fastest growing trends in digital marketing, it is ALWAYS useful!! Grab this book to gain the knowledge you need to grow your personal or professional brand!! Thanks for putting this together, Ross!"

-Marisa Cali, Live on Social Now Productions

"Ross Brand has done it again. At a time when digital platforms change by the minute, Ross Brand has once again collected insights from feet-on-the-street professionals sure to increase your influence and help your business grow. I enjoyed the first version of Ross' collection immensely. And yet, somehow, he's topped himself with Volume 2. Well done, Ross and team!"

-Mark Babbitt, co-author of Good Comes First

"As a business owner, I'm always looking for ways to grow my business. This book offers practical and helpful suggestions for succeeding in today's market. I found it helpful to have insights from so many experts all in one place!"

-Paul A. DiMilla, Amazon Customer

"Ross Brand's 100 Livestreaming & Digital Media Predictions is a must-read for any business owner of any size business who wants to grow and stay relevant. This book is not just Ross sitting down and coming up with 100 predictions. This book is filled with predictions from the top 100 online gurus. People like Winnie Sun, Dale L. Roberts, Jennifer Quinn, and Nick Nimmin have provided Ross with thoughtful predictions for 2022. Dale L. Roberts nails it in the Forward when he says Ross Brand is the Kevin Bacon of livestreaming and digital marketing. Ross has six degrees of separation from literally everyone who has livestreamed or done any digital marketing online. Ross is the King. And, that's why he was able to get 100 of the best and the brightest to contribute to his book. Needless to say, I highly recommend Ross Brand's 100 Livestreaming & Digital Media Predictions. If I could give more than 5 stars I would."

-Brad Friedman, author of *The Small Business Owner's Guide to Inbound Marketing*

"Ross Brand has done it again. It is no easy feat to gather information from so many industry professionals and put them together in one place. This is my second year purchasing this book and the insights are always beneficial to my live streaming and to my business."

-Chad and Lenyi Illa-Petersen, The Story Catcher

"Take your business to the next level with 100 digital media & livestreaming predictions! The digital media and livestreaming landscape is changing by the month. How does a business owner stay on top of the most effective approaches and best platforms in this hectic climate? By reading #100Predictions, Ross Brand's second volume in this important series. Learn from Winnie Sun, Mark S. Babbitt, Tim McDonald, Phil Gerbyshak, Bryan Kramer, yours truly, and many more. Highly recommended!"

-Chris Edmonds, co-author of *Good Comes First*

Ross Brand's second book brings with it a plethora of excellent content from high profile and well-respected industry professionals. Delve into pages packed with nuggets that inform you, prepare you, challenge you and pique your curiosity about a predictable (or not) future for livestreamers and digital media content creators. Ross is one of the most highly respected people within the livestreaming industry. Just look at the talent who are delighted to be asked for their educated opinion. I highly recommend purchasing and reading this superb book.

-Dr. Stuart Buchan, StuBu

"All you need to meet the future with confidence! So many great predictions by so many people of different disciplines. It's fascinating to see everyone's thoughts on the future of livestreaming & digital media after what we've been through over the past couple of years. I look forward to acting on some of these over the next while! Thanks for putting this together, Ross. Really well done!"

-Jodi Krangle, voice actor and audio branding expert

"Ross integrates digital marketing into this narrative and does so through the voice and lens of experts that make it their life's work. This is a must have resource guide for anyone looking to create and execute an engaged digital marketing plan."

-Amazon Customer

"This book is a great read & very informative, and should be considered one of the bibles of the live streaming world for anyone & is also a great follow up book to the 1st edition & both together are a great set & I have both. The author, Ross Brand is very knowledgeable & I consider him an expert in this field & I follow him on all socials to keep me informed."

-David Simon, Cloud Video Network

"When I saw this new release, I knew I had to add it to my collection. The book included many content creators that I am familiar with and have learned from for years. There are also creators that I look forward to exploring their recommendations for succeeding in the livestreaming space. The book is a powerhouse that is sure to have something for everyone. It will be fun to see how on point the predictions pan out."

-The Creative Lady, Amazon Customer

"Thanks Ross! As a relative newcomer to the live streaming space I consider this the Bible of livestreams. The best part though is that you have exposed me to so many giants in the space and their insights and experiences. This is a road map for the future of where livestreaming will be in the next few years. A fast paced read that I could not put down!"

-Jim Mullen, Amazon Live Influencer

"An excellent compilation of wisdom that everyone should read. Thanks Ross!"

-Amazon Customer

"I loved seeing so many perspectives and thoughts on the future of livestreaming and future-media, and not just because. I now have so many ideas I hadn't thought of to try."

-Beth Granger, CEO of Beth Granger Consulting

The ROI of learning and integrating Web3 strategies is that you will still be in business in five years!

-Jennifer Quinn aka JennyQ

Livestreaming will still be one of the best ways that brands can authentically connect with their customers.

-Nicole Sanchez

In 2023, digital marketing will rely heavily on AI technologies to provide highly tailored and personalized experiences for customers.

-Bryan Kramer

TABLE OF CONTENTS

FOREWORD BY CHRIS STONE & JIM FUHS

Ross Brand has been a trailblazer in the world of live video for many years, using his expertise and passion to connect with audiences and drive innovation in the field. From his early days as a pioneer in broadcasting, to his current role as a leading influencer and thought leader, Ross has consistently found new and creative ways to use live video to engage and inspire audiences. When you hit play on one of his videos and hear his trademark "Hey gang," you know you are about to learn something about live video. His impact has been felt far and wide, and it is with great pleasure that we introduce this book, which explores the many ways in which Ross has shaped the lives of others in the world of live video and how it will do the same for you.

"We wouldn't be here if it weren't for Ross Brand." We've said this in maybe a hundred or more livestreams, interviews and podcasts. This is not a statement that is based in hyperbole or meant to grab someone's attention when watching our show and/or listening. It's 100% true. From his impactful videos and livestreams that introduced us to tools and concepts to the conversations both virtually and in person, Ross has been an unintentional mentor. But this statement is not exclusive to the Dealcasters journey as content creators. Perhaps thousands of podcasters, livestreamers, influencers and content creators — many of whom are in the pages of this book — can say the same thing.

One of the most vital elements of building your brand as a content creator, author, speaker (or any business for that matter) is learning how to both ask and give help to others around you in your space as you move forward and achieve success. As you take courses, read books like this, watch videos, join groups, network with peers and develop skills, it's important to reach both back (or out) to others and share this information with them so that you can enjoy their successes as well. It's a selfless, abundance mindset that will pay you dividends as you move forward. No one embodies this concept more than Ross Brand. Case in point: this book! Ross could've sat down and utilized the many years of experience he's had, his learnings, stories, and taken full credit (as he should) for authoring a massively impactful guide for livestreaming and content creation. Instead, Ross Brand reached out (far and wide) to 100+ others who could deliver more value to you than you can imagine.

Whether you are a professional in the field of live video, or simply someone who is interested in the power of this medium to bring people together, we believe that this book will offer valuable insights and inspiration.

It is our prediction that you'll also be saying "I wouldn't be here if it weren't for Ross Brand" after reading this book. Welcome to the "gang."

Chris Stone & Jim Fuhs
-Dealcasters

PART I
THE BIG PICTURE

INTRODUCTION

Rolling into 2023, we are sitting on the cusp of major changes in the digital world. The old questions of where to stream or what microphone to buy seem quaint in comparison to considerations creators face in the coming year.

The big questions remain consistent over time: How do I leverage digital media to connect, grow and advance my business objectives? What content attracts audiences? How can I interest people in what I'm offering? What gear and software do I need to execute my vision?

What makes 2023 substantially different is we can no longer deny that the digital landscape is changing. Web3 is upon us, legacy social networks just aren't that into the creator anymore, and global economic indicators suggest hard times are coming.

Even a YouTuber with nearly 1.5 million subscribers is aware of the challenges of connecting with your community on the big social networks. "Content creators and business owners are tired of dealing with algorithms, notifications, and instability from social media platforms, limiting and restricting communications," says Justin Brown of Primal Video. "People want to own their communication pathways to their audience."

The new questions bubbling to the surface reveal trepidation that familiar ways of online engagement are being disrupted. Uncertainty abounds for creators and small business owners:

- Will AI make my role obsolete?
- Can social media still work for someone like me? Is it even social anymore?
- When everyone we know has a podcast, course and Facebook Group, it's natural to wonder — am I doing enough to cause people to want to listen, buy or join?
- Why bother to go live when the algorithms aren't notifying most of your network, only placing your content in front of a handful of people?

However, with all change comes opportunity for those who understand the landscape and are most prepared for the new reality. If you're looking for my recommendations on the latest and greatest platforms, gear and software, the Resources in the back of the book are for you.

For Volume 3, it would be author malpractice if we didn't devote the introduction to spotlighting new ways creators, including those featured in the following pages, are building community, influence and revenue in the current economy.

Vertical Video

People of good faith can debate the pros and cons of the TikTok-ification of online media until the cows come home. It won't change the fact that audiences are paying attention and platforms are prioritizing it. "Short-form vertical video will be the primary focus of all the major platforms in the creator economy in 2023," says Professor Nez, a prolific video creator.

If Facebook, Instagram and YouTube are to remain central to your digital outreach, then a glass-is-half-full approach is required regarding Instagram Reels and YouTube Shorts. And delivering appealing vertical video may just help people find the rest of your output.

As YouTube expert Andrew Kan says, "Vertical video will be just as important as (landscape) videos and streams. People will use that content to lead to other pieces of content."

Award-winning television broadcaster and YouTuber Michael Kinney concurs. "Longer-form content creators will see a boost in revenue by using YouTube Shorts to drive traffic to their long-form, monetized videos," Kinney says.

Jim Fuhs of Dealcasters Live thinks it's only a matter of time before more people are hosting livestreams in the vertical format. "Vertical video livestreaming will also continue to grow along with the use of YouTube Shorts, Reels, and TikTok," says Fuhs.

One of the people still building thriving communities on Facebook despite the algorithmic challenges is Eddie Garrison of Clover Media. He makes no bones about the importance of vertical. "Brands that don't adapt to this (short-form video) popular style of content could see their audience look to other brands to meet their needs," says Garrison.

Struggling to get traction on YouTube? Embracing vertical video might just open other doors for aspiring influencers. "With video becoming more important on the Instagram platform," Jaime Cohen of The Right Words says, "creators will start referring to themselves more exclusively as Instagrammers."

Event organizer Tim Gillette and live video producer Marisa Cali see short-form video coming to Twitter and LinkedIn, respectively, in the coming years.

Live Shopping

Dealcasters Live hosts Chris Stone and Jim Fuhs broke new ground on Amazon Live by bringing on compelling guests with products for sale on Amazon. They provided entertainment along with information on a platform where most live videos are product pitch fests by a solo host. Not only did this lead to revenue from sales, but it also massively raised their influence online and helped grow their existing media production and marketing businesses as people wanted the same type of dynamic show for themselves.

While Chris Stone's graphics and video production give Dealcasters Live eye-catching visuals across Amazon and social media platforms, he says it doesn't require a fancy setup for someone to have success with live shopping. "Pretty much anyone with a smartphone, some products they like, and a few thousand followers or subscribers can become a live selling influencer," Stone says.

Heil Sound is one of the brands that works with Amazon Influencers. Michelle Levitt, Heil's marketing director, says, "Amazon Live influencers are going to become a big thing in 2023, literally marketing and selling your products for you."

Livestreams focused on shopping aren't limited to Amazon Live. Hollywood insider Joie Gharrity recently launched the Superstar Women Entrepreneurs Media Network. Gharrity is bringing on women who are experts in their business niche as hosts. Selling is a part of every show.

"Livestream-Podcast will be the new reality star power vehicle to sell to niche audiences," Gharrity says. "Big brands will see the power of empowering the new realty stars because of the emotional connection they have to their niche audience."

Her network delivers tangible content that women entrepreneurs globally can implement into their personal and professional lifestyle. The videos are shared on traditional social platforms while audio versions are distributed in partnership with an established podcast network.

Don't be surprised if Instagram and Pinterest feature live shopping prominently on their platforms at some point in 2023.

Newsletters

If you're thinking, "Email newsletters? That's so 2005. Nobody wants those," then you've probably never read one of Kim Doyal's newsletters. Doyal is online entrepreneur with a track record of success as a podcaster, blogger, course creator, social media community builder, author and much more. What's a major focus of her content these days? Email newsletters.

With the #FtheHustle newsletter, Doyal starts each edition with a personal story typically based on something she recently experienced and perhaps a lesson she learned from it. She then ties it together with a project she's working on or a product her readers might find beneficial.

Doyal's newsletters also become blog posts on her website. "More people will create content on property they own (a website), instead of social first, with long-form content becoming more mainstream," she says.

In addition to creators and marketers reaching readers directly via email, newsletter platforms like Substack are drawing in well-known media personalities who are free to express themselves without fearing social media censors. Subscription-based newsletters allow you to monetize your content as well.

You don't have to be famous to issue your own newsletter on Substack. With readers signing up through email, social media algorithms become irrelevant as everyone who subscribes is then emailed when you publish new content. By the way, that content isn't just writing. Substack posts can contain podcast audio as well. The creator decides what is free and what goes behind the subscription paywall.

Both Twitter and LinkedIn have added newsletter capabilities to their platforms. Podfest founder Chris Krimitsos sees this trend picking up in 2023 and beyond. "Custom newsletters like Substack will rise over the next few years with revenue pipelines brought in to monetize these newsletters," says Krimitsos.

Jennie Mustafa-Julock (aka Coach Jennie) is someone who cares much more about building community with her clients and potential customers than loyalty to any media format or platform. She's gone all in on newsletters. In 2023, Coach Jennie says, "A massive wave of internet marketers turn away from social media and go back to basics, focusing on tools that rely less on algorithms and trends and more on content depth and value."

Artificial Intelligence (AI)

The best way to avoid being replaced by a new technology is to learn how to use it and find ways to work with it. I remember when Big Data was causing the kind of angst that AI is now. People were asking why we would need executives (who previously relied on experience and gut instinct for decision-making) if data can give us the answers. To paraphrase HR Avant-Garde president and NYU professor Vincent Suppa: Those executives who are most valued are the ones who understand what the data is recommending but are able to pinpoint those times when it is best to go against the decision recommended by the data.

With AI, the copywriter or editor won't be replaced, but their job description will change. AI can speed the time-to-market for written content by generating outlines and first drafts. The writer-editor who finds consistent work in this new AI era is the one who can spot the weaknesses in the writing such as where it is too generic, where it repeats itself and what real-world examples from research or the writer's own experience add specificity and depth to machine-produced content.

"AI is yielding its promise as the most significant technology of a lifetime. In this decade, we will witness Artificial General Intelligence in which AI will match humans in computational power," says John Pretto, an early adopter of livestreaming technology.

YouTuber Diana Gladney says creators should already be exploring how to leverage AI in their outreach and businesses. "It's extremely important for creators to identify how they can integrate AI technology to learn to better serve their community and clients," says Gladney.

The founder of the Human-to-Human (#h2h) movement, Bryan Kramer of PureMatter, sees AI tools becoming ubiquitous across digital marketing and sales applications. "AI-driven processes such as automated lead generation, dynamic pricing, and targeted content creation will become commonplace," Kramer says.

Let's not overlook extremely positive ways AI is already impacting lives. It makes our audio content more accessible through real-time transcriptions and captioning. It also enables fast turnaround for repurposing audio and video content through text-based editors such as Descript.

"By keeping an open mind and willingness to experiment with different AI tools and techniques, video editors can discover the benefits of using AI in their workflow," says YouTuber Gord Isman.

"People will combine all types of AI to create an entirely new world of content creation," adds Erin Cell of Socially Powered.

Keep an eye out for both exciting and potentially frightening aspects of AI's development. Some business owners like John Kapos of Perfection Chocolates are weighing the positives and negatives. "AI can benefit small businesses, but owners need to carefully consider the right fit — human to human — because nothing replaces wonderful, amazing, nurturing customer service," Kapos says.

However, Lights, Camera, Live host Stephanie Garcia is delighted by the potential of AI for customer engagement. "Personalized videos using AI will be a game changer. Companies will be able to call out their customers by name without any manual recording," Garcia says.

"We're talking automatic personalized video messages delivered over personalized emails and branded video pages that really drive engagement and customer loyalty," Garcia adds. "Video is already the most effective marketing and engagement content, so adding personalization to that mix is a surefire way to increase customer lifetime value."

For content consumers, Focusrite's Dan Hughley sees a better experience for the listener and viewer. "With shows now 100% controlled and programmed by AI, the sky is the limit. Expect better news and entertainment because of it."

YouTuber Dan Currier has the last word, for now, on AI. "Regardless of your current thoughts on AI, you will be hearing a lot about it and you will likely be seeing its work everywhere... but will you even know it when you see it?" says Currier.

Niche Community Platforms

What Substack is doing for newsletters, platforms like Locals and Volley are doing for community. They're giving creators direct access to their audiences free of algorithmic bias, allowing freedom of speech and monetization.

Eric Hunley is a popular livestream talk-show host on YouTube with an active community on Locals. His fans and supporters are people who enjoy his work and share a strong interest in further discussion on the topics covered on his shows. It's the kind of gathering you'd expect to find in a Facebook group, but without the distraction of everything else on Facebook, and not at the whim of Facebook's algorithm. Plus, Hunley can accept paid subscriptions for premium content, including private livestreams, in addition to the content he posts for free users.

"Creators, much like musicians of now, need to reach out to people directly," Hunley says. "We need to communicate with and talk to our audience."

Volley also enables direct access to community, but does so through asynchronous conversations or brief audio, video and text messages in threads and channels. It's like a combination of Slack and Discord but less text-based. You really feel like you are spending time with people in your community because you are seeing and hearing them participate in community conversations and direct messages without having to coordinate schedules to be online at the same time or sit through endless Zoom meetings.

Social Media News Live co-host Jeff Sieh started a space named after his show in Volley. Even social media professionals like Sieh need new outlets for community. "The major social media platforms move more toward entertainment instead of building communities," Sieh says.

Relationship marketing expert Jessika Phillips of NOW Marketing Group explains the importance of providing opportunities for direct conversations with your followers and fans outside of the major social platforms. "Relationship marketing grows momentum, turning followers into fans, clients into advocates, team members into evangelists, and community into collaborators," she says.

It's not so much about how many people you can reach or followers you can acquire on a major social platform as it is about how well you serve those smaller, but highly committed superfans interested enough to join your community and even pay to support your ability to continue creating. This means a mindset change from a focus on social media metrics to business metrics. Ten viewers with three who want to hire you is better than 1,000 views from people who watch and leave.

As Katie Brinkley of Next Step Social Communications explains, "Community will be the primary focus for creators. Not growing 'followers' but growing a tight-knit community to monetize their creativity."

LinkedIn and Pinterest

Frustrations about the Facebook algorithm discourage creators, Twitter's in flux, and Instagram's focus on Reels makes it more of an entertainment rather than social platform. Fortunately, there are other social networks that are well-known and yet have not been as fully embraced. These options for posting and engagement may see their value continue to rise in 2023.

Jon Burk leads digital marketing efforts across a wide swath of online platforms for Al Roker Entertainment. "Nobody trusts Facebook anymore. There's nobody young going on to the platform. Facebook's not growing and they've had layoffs. Twitter's destroying itself. And Instagram is competing with TikTok," Burk says, before turning a positive light on other major social media.

"There are some platforms out there that are solid as a rock, that are doing well. And one of them is LinkedIn. And the other one is

Pinterest. So those two are still the stalwarts, if you will, of social media," says Burk.

Given that they're free, it's wise to take advantage of LinkedIn and Pinterest. "That's great real estate for anybody to have as a creator," Burk says. "Spend time in those areas and really just grow your networks within those two platforms, because they are consistent, strong and trusted."

Brian Wallace of NowSourcing knows all about how to garner media attention for his infographic clients. He's long been an advocate of creators leveraging LinkedIn. "The true business social network, LinkedIn has seriously doubled down on giving proper tools and support when it comes to the creator community," Wallace says.

Meanwhile, "Tweetinggoddess" Samantha Kelly sees opportunities on LinkedIn where others may not be looking. "LinkedIn Audio is probably one of the most underused and powerful things out there for building your personal brand and an engaged community," Kelly says.

Judi Fox, who runs the LinkedIn Business Accelerator, agrees. "LinkedIn will become an even more powerful platform for podcasters and social audio events to leverage," Fox says.

Courses

Another powerful opportunity for creators in 2023 and beyond is to monetize their knowledge and expertise. Many people are reconsidering college and post-graduate education, opting instead for less expensive ways to acquire the skills needed to land jobs or thrive as entrepreneurs.

Employers and clients are less interested in the prestige of the degree and more impressed by how quickly, efficiently and successfully the job gets done.

With demand for self-directed and affordable online education growing rapidly, who is better positioned to facilitate that learning, either as instructors or producers, than digital creators with audio and video skills?

Whether it's sharing your expertise as an instructor delivered through on-demand videos and live Q&A's or helping educators transitioning to online acquire gear, set up software, and record or livestream their content, digital creators have the potential to gain income in the edtech space.

"Fewer people are choosing to attend in-person classes, workshops and learning institutions," says Dorien Morin-van Dam of More In Media. "Why would they – when time and money can be saved by learning from anyone who chooses to broadcast from anywhere in the world?"

While Morin-van Dam sees snackable short videos and social audio for "quick wins, actionable hacks and proven tips to implement instantly" as the most desirable educational content for online audiences, others are betting on bigger-ticket offerings.

In fact, it's quite possible some creators will build a full-time business from their educational products and services. Toni Henderson-Mayers is a TV actor, livestreamer and author who also runs her own company. She says, "Business owners, content creators, and brand developers will offer signature courses; some will have mini academies offering a variety of learning experiences."

John Piper of Advanced Cloud Communications is committed to the online education sector in 2023 and is growing his email list so he doesn't have to rely on social media platforms to reach his target audience.

"I do believe there will be amazing growth in niche learning opportunities utilizing digital courses, live events and podcast style audio to complement the learning process," Piper says. "Technology has enabled those that have experience and can share their story in a compelling way that the student can relate to will win and grow others."

Self-publishing authority, livestreamer, podcaster and YouTuber Dale L. Roberts goes about educating his community on self-publishing in innovative ways. He launched his own streaming service for subscribers interested in learning how to publish, market and sell books.

Roberts also collaborates with brands to create "Book Rescue," an original, episodic YouTube series. He's truly using the full range of his video skill set to help others and grow new streams of revenue for his business. "The creators who embrace all aspects and iterations of video can win big," Roberts says.

Web3 & Metaverse

"The ROI of learning and integrating Web3 strategies is that you will still be in business in five years!" That is the motivating line from Jennifer Quinn of JennyQ Live. Considering she previously said the same thing first about social media and later about livestreaming, learning more about and integrating additional Web3 tools is one of my top priorities.

So, what is this Web3 anyway? For that I turn to Mitch Jackson and Garrett Jackson, authors of the new book, *The Web3, Metaverse and AI Handbook.* Helping the Jacksons with this publication is their AI personal assistant, Kai.

Fortunately, an entry in the book called, "Explain Web3 to Me Like I'm a 4th Grader," provides a good place for me to begin:

> *Web3 is a new way of using the internet that's all about connecting people and computers in a more decentralized way. Instead of having just a few big companies control everything, one of the ideas about Web3 is to have lots of different people and organizations work together and share information.*
>
> *One way Web3 does this is through something called the blockchain, which is a kind of shared record-keeping system. It helps make sure everyone can trust the information they find online and that no one can cheat or change things without everyone else knowing.*[1]

While a full exploration of Web3 is beyond the scope of *100 Predictions*, in the pages to come, our contributors share why they see Web3 technologies as essential to their 2023 outlook and how they plan to integrate them. For anyone seeking a deeper, more comprehensive dive into Web3, I highly recommend *The Web3, Metaverse and AI Handbook.*

---

[1] From *The Web3, Metaverse and AI Handbook* by Mitch Jackson & Garrett Jackson, 2022.

Virtual reality and augmented reality also are a part of this new Web3 universe. "More creators will take advantage of Web3 and the Metaverse to build community and create, share, podcast, and hold virtual events," says The Friedman Group founder Brad Friedman, host of The Digital Slice Podcast.

Live video producer Phil Kluba of Press It sees new opportunities to use AR. "Augmented reality will become a larger part of media production and broadcasting and we'll see a lot of it in sports entertainment," Kluba says.

MediaScope host Chetachi Egwu, Ph.D. aka Dr. Tachi is betting on the monetization potential of Web3. She says, "Expect to see more ad supported streaming services, more live selling apps, and a deeper dive into the financial benefits of the metaverse/Web3."

Virtual Events

Not everyone is a fan of virtual events. Perhaps you think virtual events are a poor imitation of the real thing destined to fade away now that in-person conferences and meetups are coming back.

But it's not wise to sleep on virtual events. And when they're presented skillfully as experiences, you won't want to sleep during virtual events either.

"Virtual events allow organizations to reach a larger and more diverse audience, as attendees can participate from anywhere in the world," says Katie Hornor of The Flamingo Advantage®. "They also offer unique benefits for attendees, such as not having the expense or time constraints of travel and the ability to still engage in real-time with the speakers and other attendees," she says.

These days you can find Brad Friedman working, connecting, and creating inside the metaverse. That's also where he plans to hold virtual events. "I will be livestreaming, podcasting, holding virtual events and recording video in the Metaverse," Friedman says.

"And I'm predicting I won't be the only one. There is discontent with many of our current 'go-to' digital platforms. Platforms that we don't have control over. I believe we are going to see people looking for new places to build community and create in. One of those places is the Metaverse," adds Friedman.

Not All or Nothing

By presenting these new and alternative opportunities for community engagement and monetization, I'm not suggesting you should fold up your free social accounts and disappear. Continue to post and engage where you have audiences. Get discovered in search engines. Have profiles that enhance your brand. These things all help with maintaining your network, showcasing your influence, and promoting your business.

But when it comes to how best to serve your audience, think about where you can gather the most engaged of that audience to provide them with a better experience and deliver premium offerings.

Throughout this book you'll gain from the insights of 100 creators and entrepreneurs who have remained relevant by adapting their approaches to the changing nature of social media platforms and digital marketing opportunities.

A new round of predictions is particularly relevant this year as competition in the digital space intensifies and the economy makes online influence a necessity. New and experienced creators will benefit from the wisdom shared in *100 Livestreaming & Digital Media Predictions, Volume 3.*

PART II
PREDICTIONS

1. Jon Burk

Nobody trusts Facebook anymore. There's nobody young going on to the platform. Facebook's not growing and they've had layoffs. Twitter's destroying itself. And Instagram is competing with TikTok.

But there are some platforms that are solid as a rock. One is LinkedIn. I think that LinkedIn's profile will rise. It's just going to keep on getting better.

This past year I put Al Roker on LinkedIn for the first time. Granted, he's a celebrity, but his profile is growing incredibly well and the engagement is really high. And I realized that this is what I should use now to market our production company's shows and services as opposed to using our company page.

Because he has a production company, I can leverage Al's name to really begin to exploit the company assets as well as do marketing on behalf of our clients. That's a selling point. It's much better for me to tell clients that it's on LinkedIn as opposed to Twitter and Facebook.

The other platform that's really doing well is Pinterest. Those two – LinkedIn and Pinterest – are still the stalwarts, if you will, of social media with great real estate for anybody to have as a creator. Spend time in those areas and really just grow your networks within those two platforms, because they are consistent, strong and trusted. And trust is one of the key things right now as people are staying away from platforms where you can't trust what they're doing with your data.

Jon Burk is the director of digital strategy & marketing for award-winning Al Roker Entertainment, managing the production and activation of digital content and social video for TV programs and branded entertainment produced by the company.

2. ILEANE SMITH

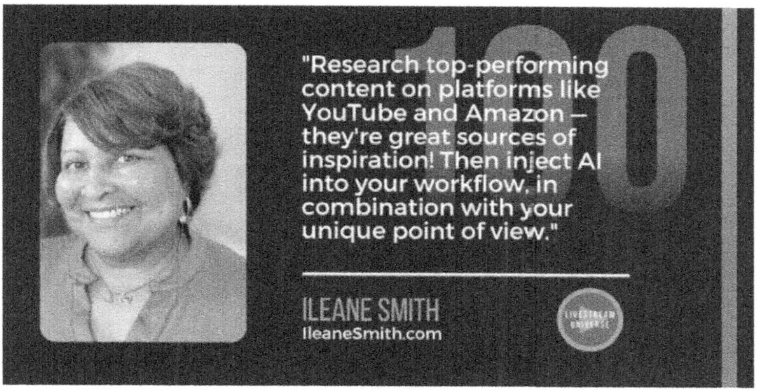

The integration of Artificial Intelligence into the creative process has removed barriers and makes creating content more accessible and easier than ever before. But it also means the next generation of creators needs guidance on how to best leverage all the content they're creating in a way that truly adds value to their audiences.

My advice? Research top-performing content on platforms like YouTube and Amazon — they're great sources of inspiration! Then inject AI into your workflow, in combination with your unique point of view to keep people engaged and coming back for more.

With a few smart strategies and optimized repurposing techniques under your belt, success is certainly within reach!

Ileane Smith is a live streamer, Amazon Influencer and content educator who has over 2.2M views on her YouTube channel.

3. Rob Greenlee

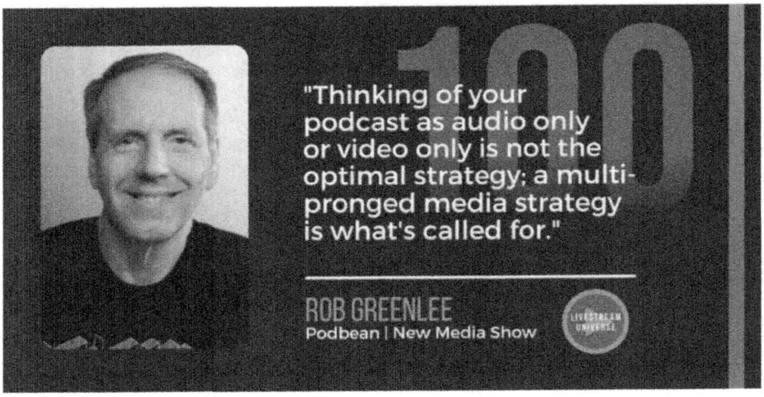

The broader scope of technology is always on the cusp of having an impact on podcasting. We're seeing that happen as AI technology and audio processing are major factors in the medium now. It's going to help with audio quality.

A big thing that's coming is the increasing use of transcripts to provide contextualization of content. That will give us a little bit deeper understanding of the content, but it also has a potential dark side too. The context at a server or computer level then can be either moderated, controlled, manipulated, or optimized for alignment with advertising.

Another big thing that's coming right now is brand safety and brand suitability. Companies that are analyzing transcripts will go out into a lot of different areas around moderation and suitability with advertisers, trying to make sure that the advertisers feel all warm and cozy about being with a particular piece of content as they move into increasingly conflict-ridden online environments.

I think the other big thing that's happening — and some of these predictions bridge the span between video and podcasting – is that video is going to have a higher place in the podcasting ecosystem regarding how people perceive what a podcast is. Not that video hasn't always been a big part of podcasting; it has, but YouTube will probably take on a new level of importance. Spotify is moving into taking a YouTube of podcasting type of a position in the medium. The disappearance of the Anchor brand and then the folding in of Spotify Creator Studios is very similar to what YouTube did. Those kinds of platforms are competing *and* aligning with each other more. It does raise the question of how Apple is going to respond.

In addition to those big trend lines, there's obviously the increasing importance of monetization and more automated advertising, which has been around the video side for a very long time. It's coming into the audio space. Maybe we're more thoughtful about how we take on advertising in podcasting because of our roots as a more trusted medium for advertisers.

Thinking of your podcast as a multimedia property is probably the correct way to come into podcasting. The days of thinking of your podcast as audio only or video only is maybe not the optimal strategy.

That's the big takeaway — increasingly a multi-pronged media strategy is what's called for.

It's also more challenging for creators to be able to navigate that effectively along with escalating expectations around professionalism. Yet at the end of the day — and you see this all the time in podcasting — it's not about how polished you are. It's about the content and how you connect with your audiences.

And that's the last thing I wanted to emphasize – how you talk to your audience and how they talk back to you is going to be increasingly important.

Rob Greenlee is the SVP of Content and Partnerships at Podbean.com, former VP of Content and Partnerships at Libsyn and cohost of NewMediaShow.com. An 18-year podcaster, Rob was inducted into the Podcast Hall of Fame in 2017.

4. Valerie Morris

Podcasting is going to become a more dominant force.

As AI and automation become stronger players in content creation, platforms like live video and podcasting will only become stronger because YOU can spotlight your personality and expertise.

You have a chance to authentically put your story and what makes your brand unique out there to the world in a natural way, and one that your followers will thank you for.

Valerie Morris is a content creator and founder of Tintero Creative, a digital marketing agency focused on content marketing and social media marketing. She is a writer, podcaster and lover of all things content online.

5 DENNIS YU

I predict that social will drive SEO.

The AI-driven bots and Google's crackdown on unhelpful content means that authentic content must come from business owners and their customers.

Businesses are the only ones with the know-how to be able to share their expertise on cell phones, so they must be the source of content.

This pushes agencies out – who can no longer rely upon thin content and auto-generated templates to drive leads. Agencies are struggling to be able to create and process video for clients, as well as being able to repurpose video to multiple social networks and eventually into blog posts.

Because Google is already looking at user signals and traffic for SEO, whether you look at CTR (usually a black hat technique to generate traffic on web pages via bots), true social engagement that flows to the web has the highest trust.

People have been declaring SEO as dead for years, but finally we see the shift from text to video – a chasm that most agencies cannot cross.

Brands will have to take content creation in-house – since UGC (user-generated content) is what powers their own content factories. They must hire dedicated virtual assistants (VAs) to process this growing mountain of content.

Thus, the shift is from high-priced agencies that are charging mainly for consulting to low-priced VAs who do implementation.

And the related shift from black box mystery to well-documented processes – like from voodoo doctors to MRIs.

TikTok is the "straw that broke the camel's back" since it finally took away the domination of Facebook and revealed that young adults do more search in TikTok than Google – and certainly more discovery.

With more searching occurring in social, true SEO now means we have to influence search results within Twitter (given new life by Elon Musk), TikTok (on a rocket ship trajectory despite Chinese ownership), and Instagram Reels (a TikTok clone).

While websites won't be the yellow pages of 2023, I expect to see a lot more traffic within apps instead of the open web. And these are social apps on ever-fancier phones, at least in Western countries.

Search isn't dead, but old SEO techniques and reliance upon Google won't be as strong. In 2024, you can tell me if I'm wrong.

Dennis Yu is the "million jobs guy" – on a mission to create a million jobs in Pakistan, the Philippines and other countries. He was an early Yahoo search engineer who has also spent a billion dollars on social media ads.

6. JUSTIN BROWN

My prediction for 2023 is that we'll see a lot of people wanting to own their communication pathways to their audience. Content creators and business owners are tired of dealing with algorithms, notifications, and instability from social media platforms, limiting and restricting communications.

This will mean a shift towards focusing on building email lists off the back of social media platforms. Even though email might seem like an old or outdated technology, it's the best way to have a direct line of communication with your audience so this is where content creators will gravitate.

Justin Brown is a Netflix-featured producer and video marketing specialist helping entrepreneurs leverage the power of online video. After 20+ years in professional video, Justin created the Primal Video YouTube channel (1M+ subscribers), growing it to a 7-figure annual business.

7. Dorien Morin-van Dam

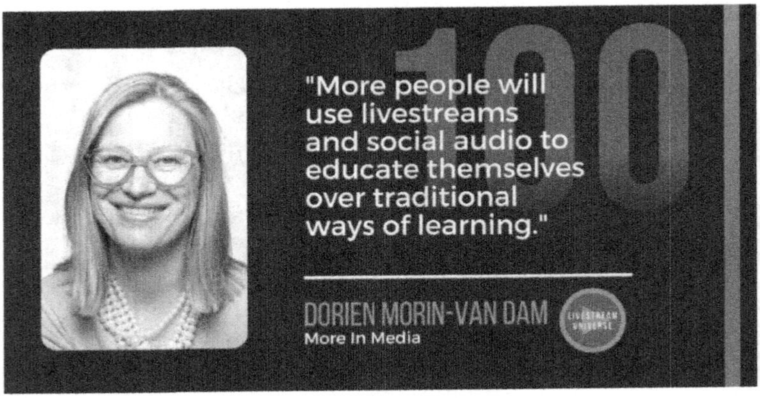

I believe that in 2023, we will see more people use livestreams and social audio to educate themselves over traditional ways of learning, for two main reasons.

1. Gone are the days of people buying full online courses and spending hours or even days on end on one topic. What people want more of are quick wins, actionable hacks, and proven tips to implement instantly! They want snack-sized mini lessons. People love short videos with stories that pack a punch. They want entertainment and innovation and, most importantly, they want to feel happy when learning.

2. Learning asynchronously is the wave of the future. With the increased popularity of remote work, newly formed hybrid teams, plus the array of available, free streaming software, fewer people are choosing to attend in-person classes, workshop and learning institutions.

Why would they when time and money can be saved by learning from anyone who chooses to broadcast from anywhere in the world?

Brand should capitalize on this trend by using livestreaming and podcasting to educate and entertain aka edutain cold and warm audiences.

Simultaneously, launch an online community where an audience can meet 'n greet and mingle with peers and where you, the brand, can foster an audience from new explorer to raving fan and brand ambassador.

Lastly, stream and podcast in the places where your audience is — and where they are most likely to go in the near future.

Do not try to build an audience in a place where you are most comfortable; instead let your intended audience guide the placement of your content.

Dorien Morin-van Dam is an Organic Specialist, International Speaker and Certified Agile Marketer at More In Media. She hosts Strategy Talks, a livestream show and podcast. You'll recognize her on stage and online by her always present orange glasses, a nod to her Dutch heritage.

8. Michael Kinney

Looking at the immediate future of online content creation, three key words come to mind: short-form, community, and authenticity.

Vertical, short-form video has been gaining popularity over the last few years, but I think YouTube's move to offer ad-based monetization on YouTube Shorts will propel short-form, vertical content creation into the stratosphere. More and more creators will be making a living on YouTube exclusively through YouTube Shorts.

And now that Shorts are integrated into the larger YouTube ecosystem, traditional, longer-form content creators will also see a boost in revenue - using YouTube Shorts to drive traffic to their long-form, monetized videos.

Livestreaming will continue to grow, but not as a vehicle for delivering content. Viewers don't really come to livestreams to learn or be informed. Pre-produced video-on-demand is far more effective at delivering informational content.

Viewers come to livestreams for interaction, connection, the excitement of a shared experience, and a sense of community. Regularly scheduled "just-chatting-style" livestreams will become an increasingly important ingredient in every creator's video content strategy.

Machine Learning and Artificial Intelligence already drive the engine behind many, if not all, social media and content-sharing platforms today. But AI is no longer just the infrastructure of a platform, now it's also creating the content. AI-generated blog posts and YouTube scripts are quickly becoming the norm.

AI video creation platforms, like Synthesia, are literally replacing traditional "human" video presenters with synthetic avatars that are digital twins of real actors. It's uncanny, but you can still tell the difference. But that will become increasingly difficult as the AI learns to act more "human." AI-generated content is already flooding platforms like YouTube, in both video and audio format.

To stand out from AI-generated content, creators will need to lean heavily on authenticity.

We'll see a move away from overly polished production toward more real, in-the-moment presentations, warts and all. Your ability to project your personality and your humanity will be the asset that captivates and engages an audience.

Michael Kinney is a YouTube Creator, Livestreamer, Podcaster, Freelance Video Producer, Occasional Actor and Former TV Broadcaster.

9. DALE L. ROBERTS

Everyone can agree that short-form video content dominates online attention these days. The recent explosion of shorter videos through TikTok leaves many industry pundits and outsiders scratching their heads, wondering if long-form video content is still relevant in 2023.

Viewers are hungrier now than ever. Twitch, YouTube, and Facebook Live are all great examples of how people still crave long-form video content—whether pre-recorded or live. The creators who embrace all aspects and iterations of video can win big.

No, video creators won't have to be a one-trick ponies — long, highly produced videos or video shorts. Now, they can paint outside the lines and produce videos without restrictions.

YouTube is a prime mover in the short-form video content arena. While the system isn't perfect and requires a lot of improvement, the YouTube development team are cross-recommending short-form viewers long-form content. And I'm sure we could assume vice versa. Ever since, more YouTube creators feel at ease with their decision to embrace all video content in all its forms.

We have a unique opportunity right now. Those video creators who adopt all forms of video content will win big in 2023. Should they spread themselves too thin? No. But, when brainstorming ideas for video publishing, they are no longer limited to one-minute or over eight-minute video content.

Why not try them all?

See what works for you, then adjust your course accordingly.

Dale L. Roberts is an award-winning author, podcaster and video content creator. He amassed over 98,000 followers across multiple platforms, including YouTube, Facebook, Twitch, Twitch, LinkedIn and more.

10. Joie Gharrity

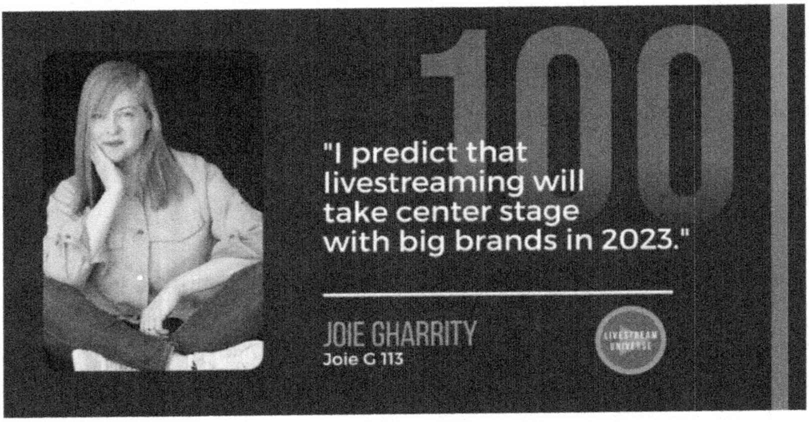

I predict that livestreaming will take center stage with big brands in 2023. They will recognize it as a resource that entertains the audience and doubles as a powerful selling tool.

Livestreaming content creators will become the new "reality stars." The marketplace will crave real-time content shared by show hosts being their authentic selves.

Just like with Reality TV, audiences are attracted to the potential for raw emotions as the hosts and featured guests let their guard down.

The marketplace wants more emotional connection on a deeper level and livestream delivers just that.

Joie Gharrity, Founder of Joie G 113, is a Brand Consultant, Personal Development Coach, Author, Livestream Host, Podcaster and International Speaker. She recently launched Superstar Women Entrepreneurs Media Network.

The Media Network empowers women entrepreneurs to create a beautiful personal and professional lifestyle with ease and grace. The media network has been picked up by a Los Angeles based agency, ElectraCast Media.

11. ERIC HUNLEY

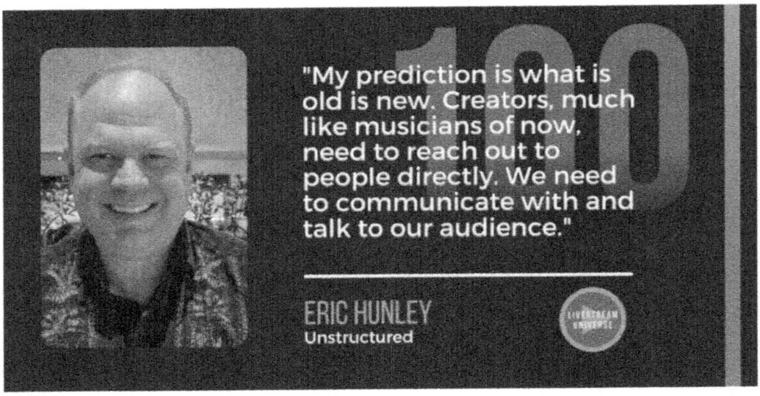

This has to be the most interesting year coming up of all. And part of it is because I don't know what's going to happen. Things are changing and they're in a flux.

We literally had a *deus ex machina* with Elon Musk buying Twitter. Who could see that coming? Who could see, at the time, the richest man in the world buying a platform because he didn't like it and taking it private? So, we are in a definite state of flux. Things can change on a dime.

My prediction is that what is old is new. By that, I mean creators, much like musicians of now, need to reach out to people directly. We need to communicate with and talk to our audience. The economy is headed to dirt. And I feel that audience communication and support – direct support – is the future.

And it really has always been the baseline. You have people like Sam Harris and others. You have alternative platforms like Locals and Rumble.

While everything is uncertain in terms of where you're going to put your content, the only thing you can truly offer is yourself and your truthfulness to your audience. Your audience will determine whether you're succeeding or failing, or otherwise. I think that's how it should be. It's not about us and it never has been. It's about what we can offer our audience. They will determine whether we live, or we fail – what we do. We will put out the content. If they enjoy it, they will support us.

And for specifics, it will be platforms such as Patreon or Locals or Rumble, or our audience will buy merch and the products that we pitch. There was an essay written in 2008 called 1000 true fans. That hasn't changed. It should always be our focus.

Eric Hunley is the host of Unstructured and the eponymous YouTube Channel Eric Hunley where he interviews a WIDE variety of folks from body language experts to YouTube lawyers, to folks in 3 letter agencies. He also has America's Untold Stories with Eric Hunley and Mark Groubert, which explores history and pop-culture that is often unknown, obscure or overlooked.

12. JennyQ

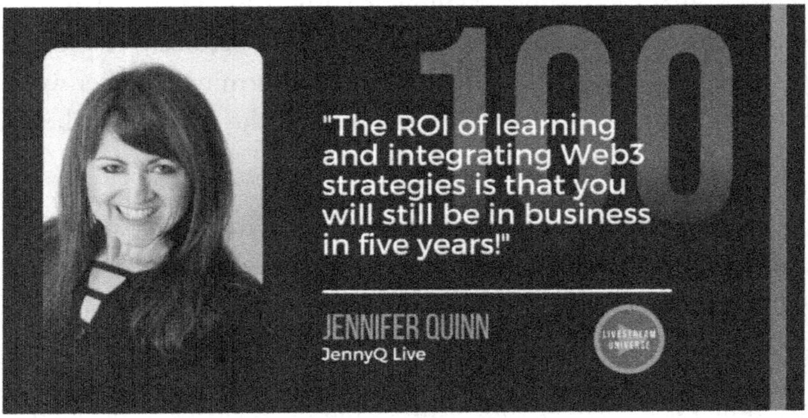

Regardless of industry or profession, when it comes to marketing, nothing will be more important this year than learning about Web3 and how to incorporate it into your business model and marketing strategy.

In 2011, 3+ years after I joined Twitter and Facebook, business owners asked the same question. Time and time again, when I would do a presentation on social media, I would be asked, "What's the ROI of social media?" I would tell them, "The ROI is you will still be in business in five years!"

Again, in 2016, just as Facebook Live was being rolled out to the general public, again I was asked, "What's the ROI of video and livestreaming?" My answer was the same. "The ROI is you will still be in business in five years!"

And now, in 2023, with even more of an urgency, because of how the entire internet of things will be evolving at a rapid pace, I say that the ROI of learning and integrating Web3 strategies is that you will still be in business in 5 years!

Jennifer Quinn "JennyQ" has been a digital marketing pioneer on all fronts. As the host of "Facebook Live's First Variety Show" to the launch of the "Being Seen" podcast, her company specializes in serving professionals by using creative and compelling marketing strategies that convert.

13. Andrew Kan

I think the evolution of vertical video will be shown. We will see it less as an offshoot, especially on YouTube. With it being monetized, I think we will see how people use that content to lead to other pieces of content.

Plan on doing a stream? Make a vertical video IN topic of the stream, and people will find/redirect to that. Vertical video will be just as important as videos and streams.

People didn't believe YouTube was the future and it was. Let's not do that to vertical video.

Andrew Kan is a YouTube Certified Expert who focuses on YouTube Channel Growth and Audience Development. He's partnered with Adobe, Godox, Joby and other brands to teach affordable YouTube and Filmmaking.

14. CHETACHI EGWU, PH.D.

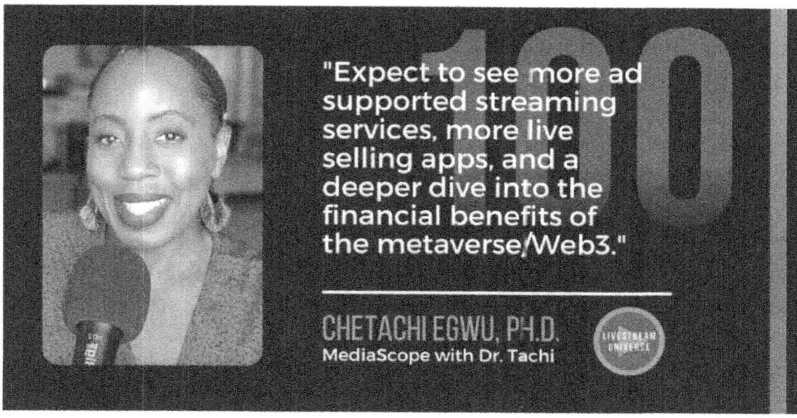

The past few years have been pivotal for media and video, in particular, streaming, and livestreaming.

From the increase in networks launching their own streaming services to the growing popularity and salience of free ad supported streaming television (FAST), it's clear that streaming is cemented as a premiere choice for entertainment and beyond.

In fact, this is demonstrated in Media Play News' report on a Q4 22 study titled "Media Advertising: Fact vs. Fiction." Fifty-seven percent of the respondents said they prefer to watch ads and pay $4 to $5 less per month for a streaming service.

Thus, I predict that even more subscription-based streaming services will debut ad supported tiers to feed the consumer need for lower cost content. With this, streamers will also realize that they will need to charge advertisers reasonable CPMs (cost per thousand) and deliver audience numbers to avoid the mistakes Netflix has made thus far with its ad supported tier.

With regard to livestreaming, I believe that this is the year that live streaming begins to "grow up." There will be more emphasis on monetizing goods and services during livestreams rather than only trying to monetize the livestream itself. Though this has been happening on a small scale for some time, I predict a steady stream of live selling platforms on the 2023 horizon.

The pandemic created an appetite for in-home services like purchasing goods, and we are not ready to let go. Additionally, we are already aware that people tend to buy from those that they know, like and trust. The engaging, real-time nature, the ability to make comments, and the excitement of bidding that some apps offer makes for a fruitful live selling experience.

If the early success of Amazon, Poshmark, and Whatnot are any indication, more apps and platforms will add and/or improve their live selling features this year.

Finally, we cannot ignore the metaverse/Web 3 as a continued force, despite consumer hesitation toward its crypto and NFT counterparts. This year, more businesses will start to include this space as part of their selling strategy instead of solely as an exploratory venture.

Hence, metaverse spaces will evolve to improve and/or implement e-commerce offerings. Many already offer the ability to stream Web 2 live streams. The next step is to also live stream metaverse-based productions for the masses to view, and for metaverse platforms to make this an easier task.

According to Forbes a recent report notes that the metaverse could potentially garner value of up to $5 trillion by 2030, a number that companies should not ignore.

In short, we can expect to see more ad supported streaming services, more live selling apps, and additions to current apps, along with a deeper dive into the financial benefits of the metaverse/Web3.

Chetachi Egwu, Ph.D. is a media professor, filmmaker, media content producer and journalist who helps people bettER their video and livestreams. She is also a crypto, NFT and metaverse enthusiast.

15. Dave Jackson

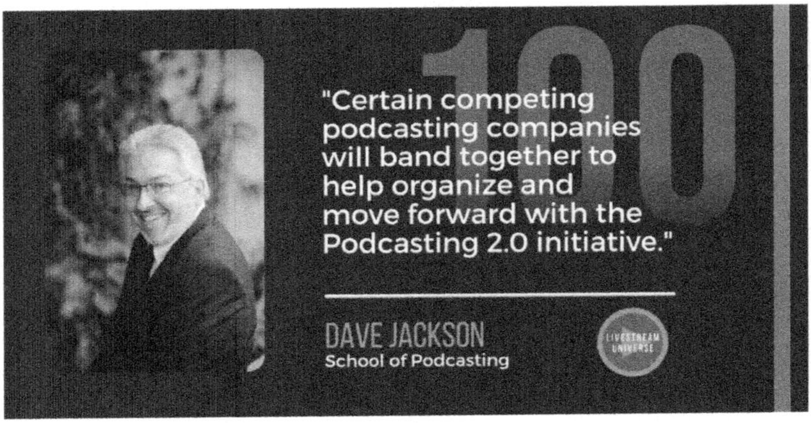

My prediction for 2023:

With more and more emphasis on being "brand safe," podcasters on networks and maybe even a few independent podcasts take out professional liability insurance to protect against claims of negligence, errors, or mistakes related to your podcast.

Often called "errors and omissions" insurance, the more income people make podcasting, the more likely they are to be sued.

Certain competing podcasting companies will band together to help organize and move forward with the Podcasting 2.0 initiative.

Dave Jackson is the founder of the School of Podcasting where he has been helping people plan, launch, grow and monetize their podcast since 2005.

16. Coach Jennie

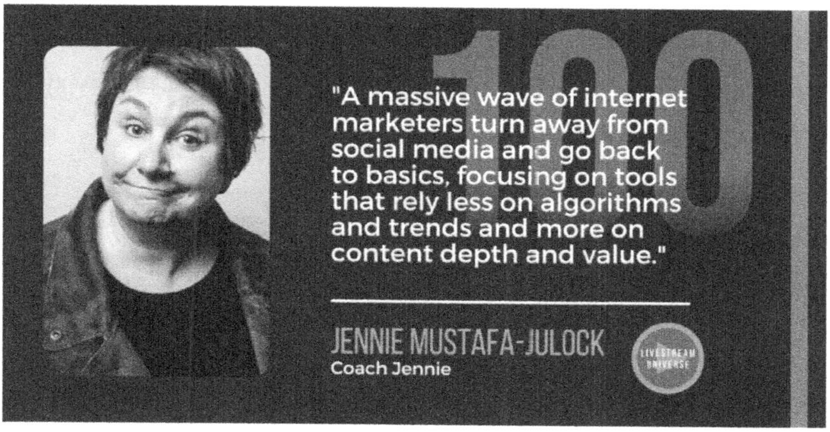

Social media has been available to the masses for more than 15 years now, and it's about time we all put down our phones.

Sure, the big players keep coming up with new features and options and algorithms (and CEOs, oof). Clearly Facebook, Instagram, Twitter, and all their cousins are not going away anytime soon, and yet another new social media tool likely arrived on the scene in the time it took you to read this sentence.

But regardless of their upgrades, social media platforms are no longer the sure thing they used to be for finding an audience and growing a thought leadership business like mine. And in today's turbulent economy, business owners must be prepared to pivot or fold.

So, I predict that 2023 will be the year when a massive wave of internet marketers turn away from social media and go back to basics, focusing on tools that rely less on ever-changing algorithms and trends and more on content depth and value. Blogging is making a big comeback these days with innovative tools - like Substack and Medium - leading the revival.

Growing your own algorithm-free blog and email list has never been more important. I, for one, am eager to ride this wave as far away from the social media monster as possible.

Jennie Mustafa-Julock (aka Coach Jennie) helps people with impatient ambition to become more productive, decisive, focused and confident so they can get on with the business of changing the world. She and her wife and CEO, Meredyth, very recently moved into their 28' Airstream trailer and are now running the Coach Jennie empire from the road full-time.

17. Ronen Artman

Having forever changed – disrupted, in fact – the newsgathering market, IP bonding is now doing the same across sports.

We've all seen over the past couple of decades the increase in the cost of many sports rights; the ever-growing demand for dynamic live content – not only around main events but all the athlete and other stories around those events; and the growth in demand for coverage of more niche sports.

What all this means is a rise in competition to bring this content to engage viewers and to successfully monetize that content. Sports broadcasting and streaming is now embracing the untethered world.

Using IP bonding brings a previously impossible level of mobility, a mobility that means filming from moving vehicles while following moving athletes is now completely possible at a highly cost-effective price point.

From mountain bike races across very difficult terrain being covered by cameras in helicopters, to high-speed motorsports, surfing to increasingly high-profile events like golf tournaments, directors have been freed from previous infrastructure constraints and can let their imaginations run wild.

All this increases fan engagement. What's crucial to this ongoing adoption of IP bonding technology across sports is the quality of the coverage, not only in its dynamic nature, but the actual quality of the images and reliability of the feed.

We all know that the sports market is the most demanding out there and this is where bonded cellular technology is essential, alongside the technical elegance of the encoding.

Then there's the ability to have multi-camera capabilities in a single unit and all at achievable price points that could only be dreamt of in the world of fiber and satellite connectivity.

We're not only talking about tier 1 and tier 2 sports; we're also talking about the huge range of niche sports that have been itching to use high-quality, live video, but have been beyond their technical and financial reach. That has forever changed courtesy of IP bonding.

As well as contribution, we're also talking about the IP distribution of that content. Matrix next-gen IP cloud video solutions allows customers to share and receive high-quality, low-latency live feeds with, and from, broadcasters and other stakeholders around the globe – inside or outside your organization.

Volleyball World is a prime customer example of Matrix use for sports distribution to its customers. Indeed, so reliable is the technology that it's now being deployed as a disaster recovery solution as requirements to ensure successful sports coverage increase.

Ronen Artman is the global vice-president of marketing at LiveU, the broadcast industry standard for remote livestreaming video transmission via patented cellular bonding technology.

18. Judi Fox

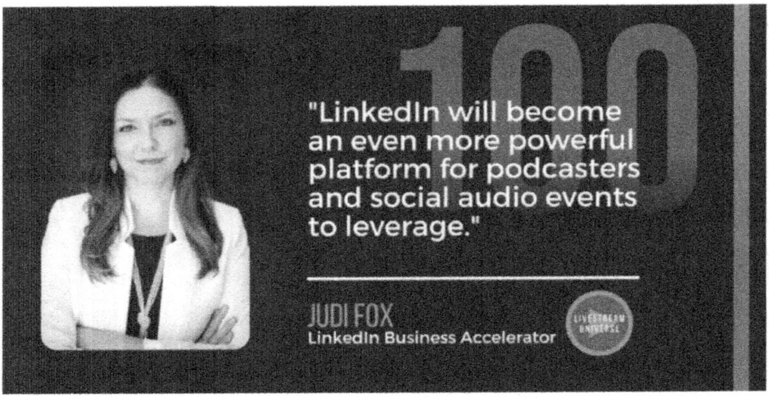

In 2023, LinkedIn will become an even more powerful platform for podcasters and social audio events to leverage.

For example, LinkedIn has already launched their podcast network and announced over 15 shows.

Plus, with LinkedIn Audio Events rolling out to company pages, I am predicting that LinkedIn will add the ability for events to sell tickets or create private invite-only events on the platform.

Therefore, a successful 2023 Marketing Strategy will benefit from leveraging LinkedIn's new tools, including:

1) Drafting and Scheduling LinkedIn Posts

During 2023 we will be able to create and schedule content natively to post on the LinkedIn Platform.

2) Embedding content in LinkedIn Newsletters

For example, we can already embed the Spotify podcast player. LinkedIn will roll out more tools and ways to access podcast episodes and social audio content directly on LinkedIn.

3) LinkedIn Audio Events

LinkedIn will roll out Audio Events to companies, which will expand towards creating ways to host private events and ticketing directly on LinkedIn.

Judi Fox is listed as a Top 10 LinkedIn Coach in Yahoo Finance. She is featured in Inc Magazine, Vidsummit, Social Media Marketing World, Video Marketing World, People of Video, and on several Top 100 Marketing Podcasts. With over 18 years of experience in business development and sales, she developed the LinkedIn Business Accelerator Method built from launching an Engineering Consulting business on LinkedIn. Clients have been implementing the LinkedIn Business Accelerator sustainable method for 5+ years achieving more business and opportunities, including adding a million dollars to a service-based business in 6 months.

19. Diana Gladney

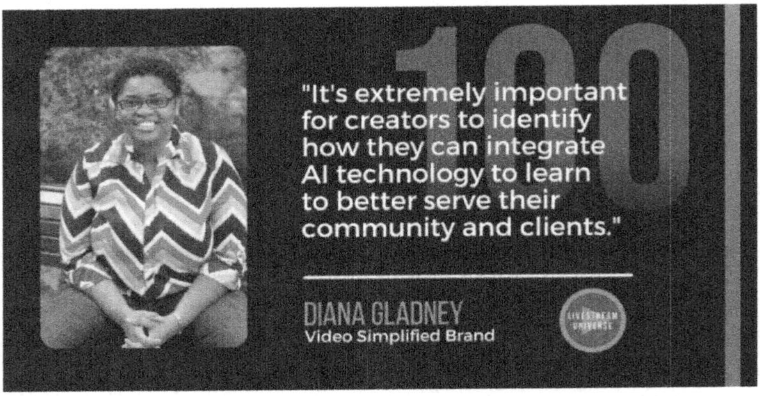

For content creating entrepreneurs, the new advancements of AI (artificial intelligence) technology will cause a new dawn of rapid information our ideal viewers and clients will now have access to that will change how they consume content.

This will require video content creators to embrace the changes and lean into a new age of machine learning technology that our ideal clients and customer are likely to use on a daily basis.

As many software tools, brands, and social platforms have already begun to integrate this AI technology, I believe it'll be extremely important for creators to identify how they can integrate this technology to learn to better serve their community and clients.

Those who may feel this new technology is a fad will miss an opportunity to more deeply connect with their audience and solidify their place as their target audience's go-to resource in their niche.

I believe all the major social platforms will find unique ways to integrate this technology to create a more enjoyable viewing experience for viewers and keep them on the platform longer.

I believe wise content creating entrepreneurs should begin to question how they can integrate this tech to better serve their people.

Diana Gladney helps entrepreneurs simplify the video creation process to amplify their brand using video on YouTube. She's the author of "The One Right Video" and host of the Diana Gladney YouTube channel.

20. Bryan Kramer

In 2023, digital marketing will rely heavily on AI technologies to provide highly tailored and personalized experiences for customers.

AI-driven processes such as automated lead generation, dynamic pricing, and targeted content creation will become commonplace.

Through AI-driven optimization, businesses will be able to create more efficient campaigns that maximize the impact of their marketing spend.

Bryan Kramer is a Proud Father of Human-to-Human (H2H). Executive Coach, Keynote Speaker, Investor, Virtual & In-Person Global Keynote Speaker, CEO of 2 Companies and TEDTalker.

21. Jodi Krangle

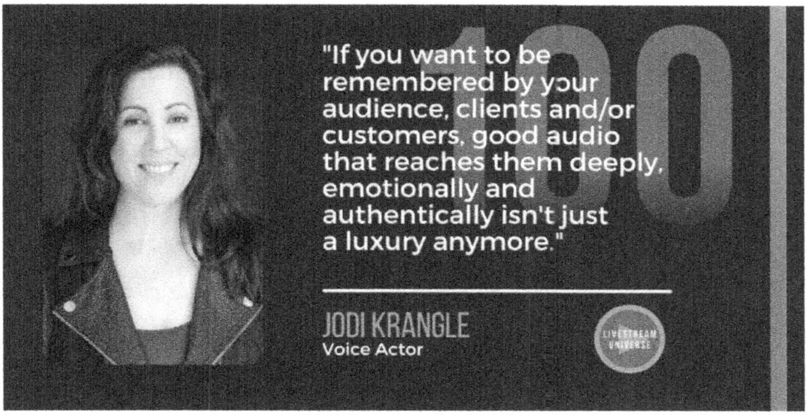

Sound will continue to be very important in 2023 - even more than it was in the previous year.

The pandemic highlighted how much we crave human connection, even if we can't have it in person. Good sound - both in livestreams, podcasts, and video/film - has made it possible for us to connect with one another authentically, even if we're far apart.

Brands are starting to realize that a comprehensive discussion about how they SOUND, is as important as what they look like. It's not just about a sonic logo, either. It's about sound that follows through with a company's DNA, stays true to it, and is heard in multiple touchpoints (on hold, commercials, public spaces, internal presentations, speaking engagements, etc.)

And with devices like Google Home and Alexa gaining more and more traction, differentiating themselves in a world that only contains audio hasn't just become preferred, it's become essential.

Podcasts are also becoming more and more popular - not just for individuals, but also for brands and companies. And an audience isn't going to listen to a podcast (or livestream) that doesn't have good audio.

If you want to be remembered by your audience, clients and/or customers, good audio that reaches them deeply, emotionally, and authentically isn't just a luxury anymore. It's a necessity.

Jodi Krangle is a voice actor of over 15 years who works with major brands all over the world including Dell, BBVA, HGTV, Nespresso and Kraft. She's been heard in thousands of your favorite commercials and on your favorite brand websites. On her podcast "Audio Branding," she and her guests offer advice on making an impact with sound along with how sound influences us, both in our buying decisions and our daily lives.

22. Chris Krimitsos

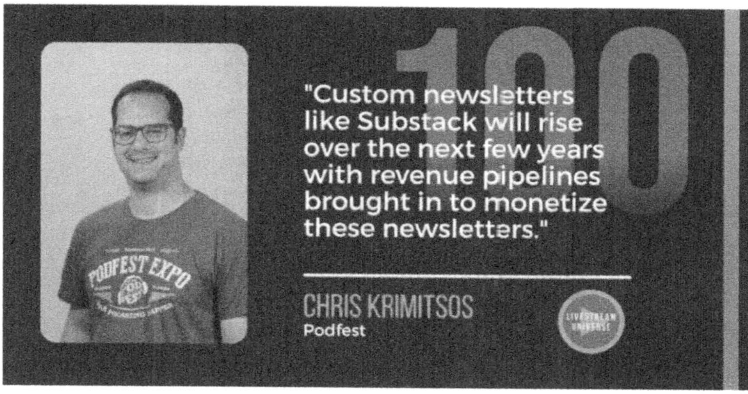

We will see custom newsletters that look like Substack rise over the next few years.

We will also see revenue pipelines be brought in to monetize these newsletters.

We will also see podcasters add streaming at a rapid pace to their shows. Monetization will grow through the Span (Spotify Ad Network) and Ad Whizz to the majority of podcasters.

Chris Krimitsos organizes a conference, Podfest, that supports the endeavors of thousands of podcasters.

23. Samantha Kelly

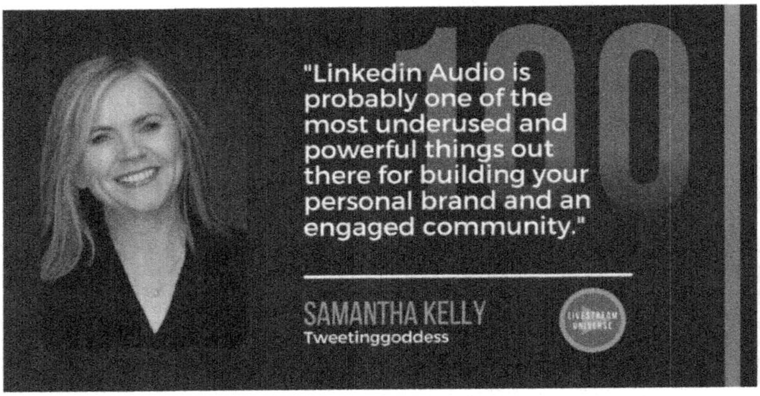

Social audio I feel will start to become more popular and be used to build trust and communities.

LinkedIn Audio is probably one of the most underused and powerful things out there for building your personal brand and building a really engaged community.

Also, Twitter spaces is going to change to video with lots more features to make it the 'go-to' platform for everything. Twitter is already the 'go-to' platform for live news as it happens so together with live streaming and audio YOU can become that live news as it happens especially if you are an expert in your niche.

Samantha Kelly is a Twitter expert, community builder, host and speaker. She's been named on many global lists as one of the top B2B social media influencers.

24. Jim Fuhs

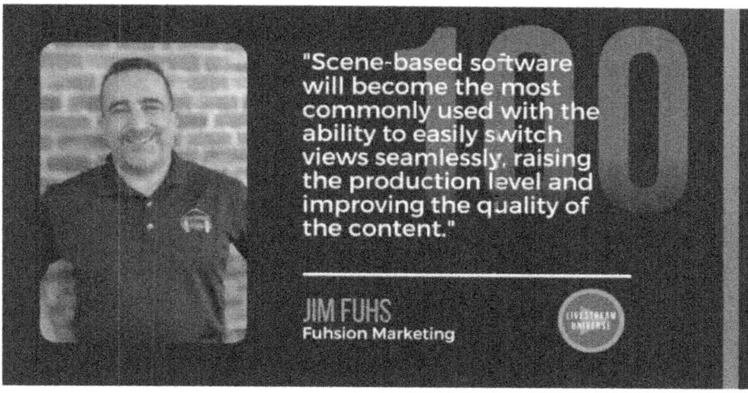

During 2023, video will continue to increase as a way for personal brands and businesses to connect with their audience.

Whether a livestream or recorded, scene-based software will become the most commonly used program. The ability to easily switch views seamlessly will raise the level of production and improve the quality of the content.

Companies like Wave.video, ECamm, and Evmux have taken the lead in this space, and it will continue to be a highly competitive space with a demand for livestream producers that can manage the programs for those who do not have the time or the ability to create scenes and the run of shows and scripts to get the message across.

Vertical video livestreaming will also continue to grow along with the use of YouTube Shorts, Reels, and TikTok. Users will ultimately want products that are easy to use but allow for full customization.

These products will also be important for virtual presentations and creating engagement with the live audience to keep their attention.

Jim Fuhs is currently the cohost of Dealcasters on Amazon, The Tim and Jim Show with over 4 years of livestream interviews, and is a remote livestream producer. Jim has worked with StreamYard, IBM Research, MIT, Carta Equity, San Francisco Chronicle and several other organizations in producing virtual events and presentations.

25. Chris Stone

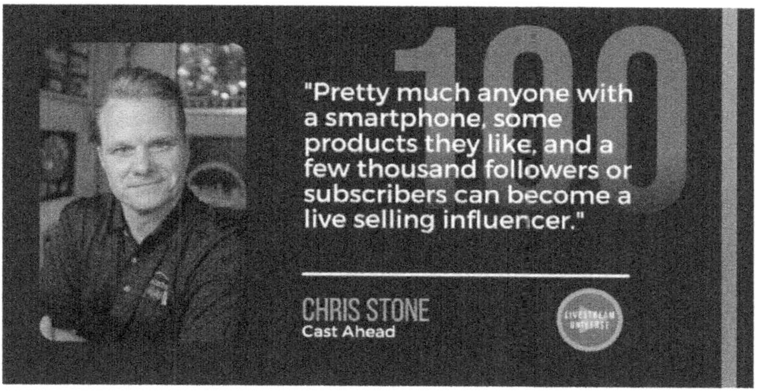

Pretty much anyone with a smartphone, some products they like, and a few thousand followers or subscribers can become a "live selling influencer." You don't even need any followers or subscribers to be a WalMart influencer.

We've seen the continued rise of the Amazon Influencer Program where people are acquiring what they consider to be "Full-Time Revenue" very quickly if they do it right.

The downside is that we're seeing a rise of these "influencers" who have never even gone live before. They are making the same mistakes that many other content creators make; the most critical mistake is they are not considering what problems they are solving for the audience who are there to be educated, entertained, and to have their problems solved. These brand new "influencers" don't know how to use simple effective tech that allows for the shoppers to hear you, lighting so they can see you, and eye contact so they can see if they can trust you in order to make that purchase.

On the brand side, companies will begin to hire people who can be a relatable and trustworthy face of their company who can go live to have a show and in a sense: become a live seller. They're also using content to build out content on YouTube and guesting on other podcasts. This gives a human face to the brand and is relatable to consumers.

Vertical shopping experiences similar to TikTok, IG Reels, YouTube Shorts— also sprinkle in the Idea Lists from Creators as well as shoppable images—will be massive.

Amazon has done it recently and people will begin to trust purchasing larger-priced items via their phones. Vertical-oriented shopping platforms are another massive opportunity for creators to create the proper, engaging content in order for the more vertical-leaning customers to make a purchase decision. How many times did we hear someone say "TikTok made me buy it?" or have bought something via Instagram directly from someone and felt duped?

Amazon already has garnered the trust of millions, Prime members get it within a few days and returns are super simple. Other sites and online sellers are launching live shopping platforms and creating influencer programs and have a way to go to compete with Amazon. But are they able to learn from some of these mistakes?

Chris Stone produces and develops shows for driven entrepreneurs who don't have the time to do it themselves. He also co-hosts the show Dealcasters on Amazon with Jim Fuhs of Fuhsion Marketing.

26. Kim Doyal

We'll see more people making sure they're also creating content on property they own (a website), instead of social first, with long-form content becoming more mainstream.

Creating better content consistently, backed by a contextual distribution plan (no more pushing the same content everywhere) will be key.

Kim Doyal is an entrepreneur, coach, speaker and podcaster. She's the founder of Email Insiders, #FtheHUSTLE and "Create It — The Content Planner." Kim built her lifestyle business through content marketing, podcasting and email marketing.

27. JEFFREY FITZGERALD

With the rapid adoption of AI for digital creation, 2023 will be the year livestream content merges with the artificial.

Artificial Intelligence continues to make significant inroads into all areas of digital content creation. From the now familiar and mind-bending art that depicts your selfie in heroic landscapes to apps that create better images from pretty bad pictures.

One thing is certain. AI is emerging as a core tool of the future for online media producers. And for a growing number of creators, AI means more than just artificial intelligence.

It will mean ALL IN.

Practically speaking, the number one challenge for anyone producing content these days is time. Doing meaningful research. Finding new ideas and content directions. Managing the burden of the creative process. And even being on camera as the face of your brand.

But the promise of what we have seen so far reveals this. AI will become a trusted partner throughout the creative and production process. Albeit one without a pulse, but one that can dig, write, speak, and even present for you - quite literally.

Now, is it at the pinnacle of human perfection right now? Far from it. But with every passing bit and byte, AI is pursuing a parallel reality to our own.

And it's getting better every day. These are just a few of the creatively synthetic offerings out there now.

- AI-generated ultra-realistic human voices that can narrate in a variety of styles, languages, and dialects.

- AI copywriting and marketing content for the web, social media, advertising, and a variety of other uses.

- AI video and podcast editing that can remove filler words like "um, "uh," and "you know" from your recordings. Even adding or replacing words in your own voice.

- AI video presentation services that can produce videos featuring a virtual you, talking as an avatar from just an image.

And while we have hardly seen the most significant contributions from AI, one thing is becoming clear. 2023 will be a paradigm shift for what we have known and seen in the content creation process.

As more of us explore this strange but fascinating new world with our digital doppelganger, how far will it go?

Right now, AI is in its infancy. But virtual presenters for sales and webinar videos are already emerging.

Though livestreaming is most likely the final frontier to fully synthesize. Because, well, it's live and not pre-recorded. *And with a very human connection to a live audience.*

But that's not to say it can't happen.

Yes, there currently are filters to make you look like a puppy while pitching your wares on Amazon Live.

But what about a true, 100% AI-generated digital version of yourself live? (If that is even such a thing!)

Imagine this version. With your personality traits. Your voice and appearance. Yet with humanly undetectable digital attributes that create targeted and meaningful marketing messages without the heavy lift. Nuanced for unique and diverse audiences. In a multitude of languages, all streamed at the same time.

Ready to 'Go Live' without actually being live... or even alive?

It's not impossible, and in fact, it's probable.

All of this raises new and serious questions regarding the social ethics around what AI might demand from us.

Answers are elusive right now as we gorge on the novelty. But the more immersed and accustomed we become to a Hallmark digital romance; one question will always remain.

'The more you rely on AI to create your best you, how much of the real you is actually left?' It will certainly be the conversation of artificial authenticity.

And like sands through the hourglass, so are the days of our digital lives.

Jeffrey Fitzgerald (JeffInHighDef) is based in the Washington DC area, and is the content creator behind Evening Star Creative. He is a three-time 2020/2022 Emmy Award winning video producer, with over 40 years in broadcast, online video and livestreaming. His work includes national television, digital marketing and first responder leadership and recruiting. Jeff also hosts The Digital Green Room, a private Facebook community for creatives, artists and entrepreneurs.

28. Toni Henderson-Mayers

Online courses will be on the rise. Business owners, content creators, and brand developers will offer signature courses, and some will have their own mini academies for customers and followers to choose from a variety of learning experiences.

Toni Henderson-Mayers livestreamed concepts from her book Wise Courtship, which obtained worldwide recognition and awards. She's developed several businesses and a successful acting career on stage and screen.

29. DAN CURRIER

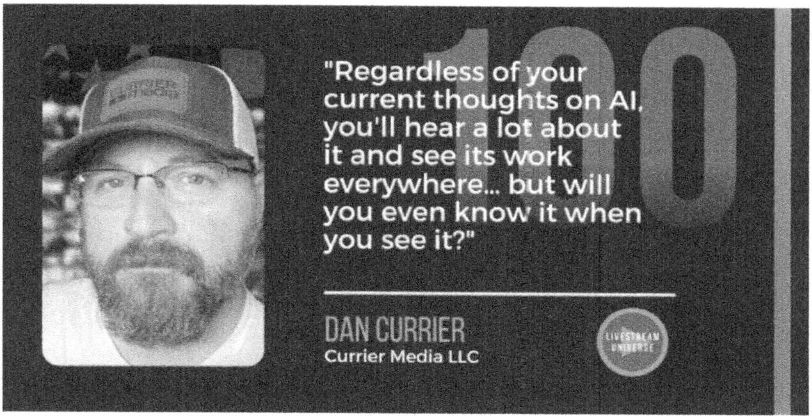

As creators and business owners, we are always looking for ways to get more done and be more efficient doing it. Time is so much more valuable than money, but money doesn't hurt either. As we grow, we are constantly looking for way to maximize both.

There is a shiny new technology that just might help us do that like never before. The emerging technology that will be on full display in 2023 is AI. Artificial Intelligence is coming into focus in a number of ways within the creator community and depending on who you ask its potential impact varies widely.

AI has already appeared in the form of "art" created by a computer that can look at your photo and create a mystical new design. It has also emerged as a powerful engine for writing new copy and content that is impressively well thought out and organized… and completed in seconds. Some see AI as a betrayal of everything creative and liken it to the fast food of the creator economy… quick and easy but lacking substance and sophistication. Others see it as the holy grail of getting more done and working smarter rather than harder.

Leveraging computer power and freeing up human hours to focus on new endeavors, may just be the ultimate in entrepreneurial go-juice. As with all new technologies, only time will tell how AI is leveraged and just how powerful it will turn out to be.

Regardless of your current thoughts on AI, you will be hearing a lot about it, and you will likely be seeing its work everywhere... but will you even know it when you see it?

Dan Currier is CEO of Currier Media LLC, which helps creators find ways to earn money with their content. He has 85k subscribers and 6.5 million views on YouTube. Dan is the Author of Influencer100x, a course helping Amazon Influencers earn thousands of dollars a month.

30. Jessika Phillips

In today's world, it is more important than ever to stay ahead of the curve when it comes to marketing. With new technologies and platforms emerging all the time, it can be hard to know which ones are worth your time and attention.

Here are digital marketing trends that you should keep an eye on in 2023. From chatbots, Web 3, to personalization and AI, relationship marketing is no longer just about the customer relationship; it's also about understanding how your business relationships and processes will evolve.

Brand culture, character, and a real commitment to doing what's right, not what's easy for customers, will prove to be a true differentiator.

Only brands that are willing to first focus on starting from the inside > out to build their foundation of momentum growth will succeed in 2023… meaning marketing is only going to amplify what's already at your core. So, if your brand core manifesto, brand culture, product, service, experience, and the process isn't the best they can be, others will notice, and marketing is only going to amplify the truth.

We can't buy community, reputation, loyalty, or trust. It must be earned and grown through building more powerful relationships than your marketing. While this takes time, it's important to remember that relationship marketing grows in strong momentum when done right, turning our followers into real fans, clients into our biggest advocates, team members into our biggest evangelists, community into our most prominent collaborators.

Marketing has evolved to us creating an authentic brand manifesto/mission and providing an immersive, personalized customer journey beyond just product promotion —beginning with our focus on developing a solid company culture that is rooted in trust and fostered through collaboration.

This is the foundation for a successful online marketing strategy in 2023. By staying on top of these trends and investing in relationship marketing strategies, businesses can stay one step ahead of the competition.

Jessika Phillips is an award-winning marketing strategist known for her outspoken dedication to relationship marketing. Jessika helps brands organically build sustainable, craveable and profitable businesses by out-CARE-ing the competition. Founder of NOW Marketing Group and the Relationship Marketing System, Jessika has been named the go-to expert on Relationship Marketing and has built a team of 16+ who serve over 100 global brands. Jessika is often asked to share her knowledge and insight as an industry-leading influencer. Jessika also gives back to her community and welcomes 300+ attendees and world-renowned speakers to her Social Media Week Lima conference annually.

31. DAN HUGHLEY

I can see more artificial intelligence being used to program shows. We now have AI that can help us with our show notes and with shows now being 100% controlled and programmed by AI, the sky is the limit.

AI is smarter and faster than humans at consuming and analyzing content as data; this tech can very quickly determine the trends and what should come next.

I believe in the next year we'll lean into that heavier and have better news and entertainment because of it.

Dan Hughley is a passionate podcaster, audio engineer and pro audio marketing professional who has worked with powerful influencers and some of the largest recording gear manufacturers in the world. In his current role as US senior marketing manager at Focusrite, Dan found his passion for podcasting and continues to create educational materials which focus directly on the needs of podcasters. Dan hosts his own podcast, The Johnny Podcast Podcast, which has received respectable preliminary reviews.

32. Caren Glasser

As more and more streaming services are hitting the airways, there's going to be a bigger need for original content, not scripted, raw, honest, thought-provoking, and entertaining.

This is what I call real reality television, and it's an amazing opportunity. As the preference for video grows, we are seeing a shift from podcast platforms.

Spotify, with around 450 million monthly average users, and 194 million premium subscribers has jumped in to offer podcast producers the option of uploading video with their audio podcast.

They had me at video, so as a first adopter, I have jumped in and now give my viewers and listeners a choice as how to consume their shows. They can listen or they can watch. I do predict we will see other platforms dip their toes into this groundbreaking opportunity.

Caren Glasser is the host and producer of 4 online talk shows, including Caren Glasser LIVE, Uncorked LIVE!, The Author's Spotlight & Minding Your Mental Health. In total, her online podcasts have amassed over 2 million views. In 2018 she was voted one of the 5 top on-line hosts in the world.

33. Jeff Sieh

In 2023 I think we'll continue to see the major social media platforms move more toward entertainment instead of building communities.

Gone are the days where friends and family you're connected with see all the content you post. The algorithms across platforms are now placing entertainment and clickbait at the top of your feed.

2023 will also be the start of the rise of Skynet! Just kidding... kinda. With the rise of written content created by AI, I believe that people will put lesser value in written posts.

BUT this is an opportunity for creators and brands to connect with their clients. AI can't yet generate a live person who can know their name, have inside jokes with because they show up each week, and answer their question on the fly about a product or service.

There has never been a greater time where people will want to make a genuine connection with a creator or brand. Live video is what can fill that need and help build a true community.

Jeff Sieh is an international speaker and visual marketing consultant. He hosts the Social Media News Live show and podcast and is also "Head Beard" at Manly Pinterest Tips. Jeff has worked with and produced a wide range of content for various companies, including Guy Kawasaki, Kim Garst, Social Media Examiner and Tailwind.

34. KATIE BRINKLEY

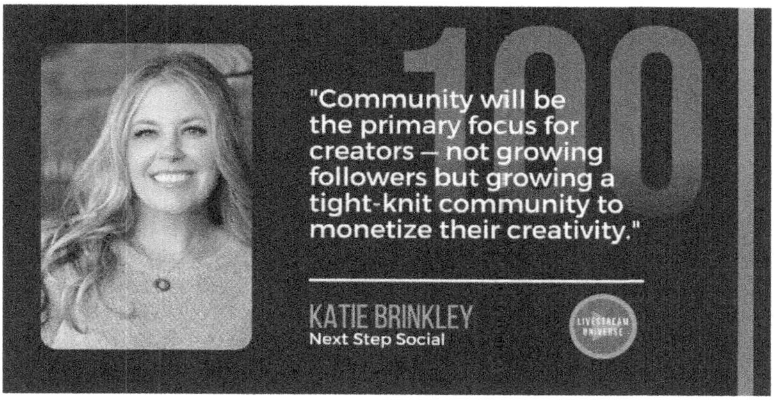

I believe that Community will be the primary focus for creators. Not growing "followers" but growing a tight-knit community from social media.

Taking a more intimate approach with your biggest fans and nurturing that audience through email, Discord, podcasting, or community platforms like Mighty Networks is the way.

Creators are tired of building audiences on "rented land," like most social media platforms, and not seeing the monetary rewards, so I believe many will look to take their audiences to a place where they have more power over how they can monetize their creativity.

Katie Brinkley is a Social Media Strategist and Founder of Next Step Social Communications. With nearly two decades helping brands and businesses with their social media, Katie has been at the forefront of the changes in how buyers engage on social media.

35. Mitch Jackson, Esq.

I'm a creator, entrepreneur, and for the past three decades, a lawyer. Wearing all three hats, I've never seen as much change and opportunity in the digital marketing, branding, and business worlds as I'm seeing now, in real-time, right before my eyes.

Artificial intelligence platforms like OpenAI's ChatGPT, DALL·E, MidJourney, Lensa, and others, create new opportunities, responsibilities, and liabilities.

Have fun and create. Take time to appreciate how intellectual property rights apply in the real world and these new digital spaces. At the same time, document your relationships and agreements.

Innovative creators and entrepreneurs who combine traditional web2 content creation with leveraging web3's new blockchain, smart contracts, NFTs, digital currency, and the metaverse, are those same creators and entrepreneurs positioning themselves for long-term marketing, branding, and business success.

Enjoy the journey and continue making each day your masterpiece.

Mitch Jackson is an award-winning 2009 Orange County Trial Lawyer of the Year and 2013 California Litigation Lawyer of the Year. He enjoys applying his three decades of practicing law to help add value to consumers and clients who are doing the digital dance at the intersection of law, business and technology, including web3 and the metaverse.

36. Michelle Levitt

Amazon Live influencers are going to become a big thing in 2023.

They are literally marketing and selling your products for you. What could be better than that?

Michelle Levitt is the Marketing Director for Heil Sound.

37. Mark S. Babbitt

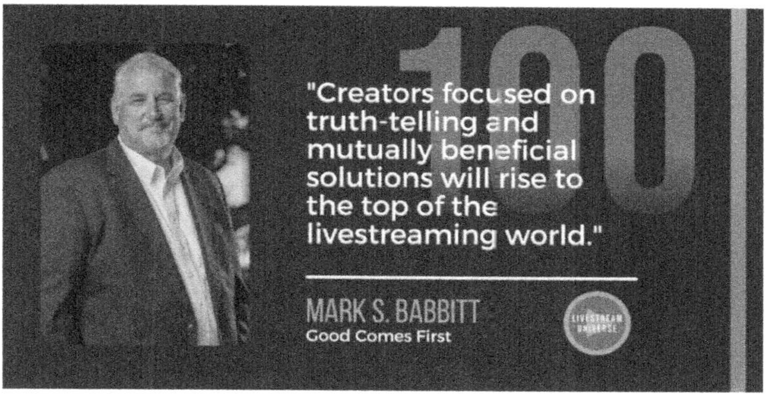

In 2023, creators will leverage digital content to create instant credibility for their brands.

More specifically, rather than take a follow-the-leader approach to generating controversy or spouting self-serving (and self-promoting) opinions, they'll deliberately lean in on facts from reliable sources to create more substantive conversations.

With the 2022 election cycle behind us and the 2024 election right in front of us, we'll earn the attention of digital consumers by caring enough to present objective data, working hard to end the divisiveness so prevalent in society today, and inspiring others to seek out the truth.

Ultimately, the creators focused on truth-telling and mutually beneficial solutions will rise to the top of the livestreaming world.

Mark Babbitt is a speaker, author, culture architect, executive coach and career mentor. He serves as President of WorqIQ and CEO and Founder of YouTern.

Mark is the co-author of two best-selling books: "Good Comes First" and "A World Gone Social." Find his work in Inc., Entrepreneur, Forbes and more.

38. S. Chris Edmonds

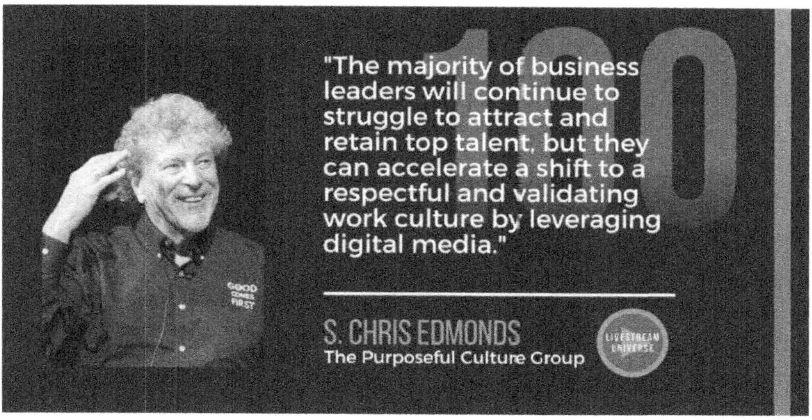

I predict that the majority of business leaders will continue to struggle to attract and retain top talent in 2023.

Far too many business leaders around the globe rely upon antiquated, autocratic, command-and-control beliefs and behaviors. These beliefs and behaviors create toxic work cultures which drive talented staff out the door, and which cause candidates to look elsewhere for job opportunities.

Business leaders have tried to attract candidates with higher pay and better benefits, but talented players are not taking the first job offer that comes their way. They're being patient and selective.

Those talented players aren't settling for a crappy work culture. They're seeking and finding workplaces where they're respected and validated for their aligned ideas, efforts, and contributions every day. Companies like Radio Flyer, ABC Supply Company, In-N-Out Burger, Five Below, and St. Jude's Children's Research Hospital attract top talent - and hire top talent - every day.

Business leaders can accelerate the shift to a respectful and validating work culture by leveraging digital media to

1) reinforce their desired culture, values, and behaviors,

2) share success stories that celebrate players and teams that model their values and behaviors, and

3) teach leaders new behaviors and new skills so they effectively praise aligned behavior and effectively coach mis-aligned behavior, daily.

Livestreaming is one of the best tools available to business leaders to quickly engage and educate remote teams and team members regularly about the critical importance of their values and behaviors.

S. Chris Edmonds is CEO of The Purposeful Culture Group, helping leaders sustain uncompromising work cultures and values. He's also a Speaker, Author, Executive Consultant and Musician.

39. Bee Smith

In 2023, the need for content in mental health advocacy is urgent. YouTube Shorts, Reels livestream video that inform, empower, and encourage will bridge the gap in truly forming communities of purpose!

Teach one, reach one.

Benedria "Bee"Smith is an international multi-bestselling author, public speaker and mental health advocate. After surviving suicide, she used her platform to tell the story of the struggle and the Overcomer, resulting in two documentary shorts, one award-winning, that highlight suicide prevention advocacy. She serves and speaks internationally to empower and inform. In 2019 she received the Dr. Murali Krisha Family award for her unique approach to Mental Health Advocacy.

40. Lee Love

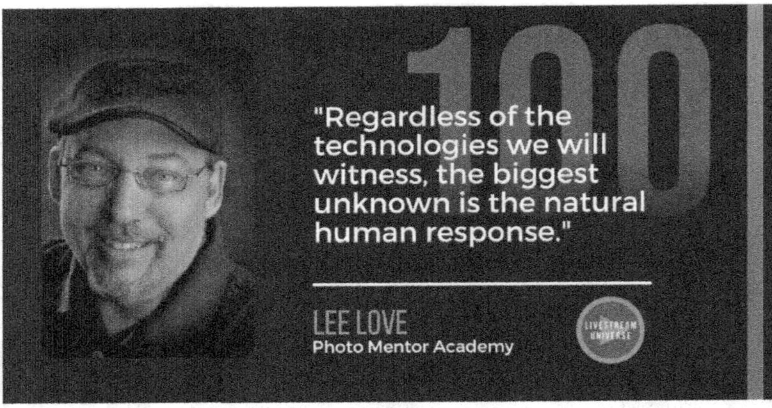

We are no doubt living in interesting times, partly due to the rapid rate of development of technology.

Since the 1950s we have seen a shorter and shorter development time frame for new technologies. For example, it has only been 16 years since Steve Jobs introduced the iPhone in 2007, and there is no question this was a catalyst for change that has affected an estimated 90.72% of the world's population.

In comparison, the Wright brothers flew in 1903 but commercial aviation was not available till the 1950s.

The question I ask is, are we able to consume and adopt technology at the pace it is now being developed? I often view many new products or services as a solution looking for a problem. This is often the case when engineers develop technology "just because they can."

By nature, we tend to resist change on many levels, which is why the phrase "In Internet Years" was used to measure technology time frames.

As ubiquitous as we consider the internet even, it did not have the current adoption rate of the smartphone. Consider that ARPANET was established in 1970, and AOL did not appear until 1989. Yet even at 19 years, the impact was not instantaneous. The next shift did not appear until my old boss Marc Andreessen announced the Mosaic browser in 1993 that the adoption started to accelerate.

Fast forward at light speed to 2023 and the same species (yea you and me) that took 50 years before being able to develop and market commercial aviation is now talking to their devices, and they are talking back! We are on the verge of self-driving vehicles and some kind of weird technology that is supposed to mimic human or natural intelligence.

Regardless of the technologies we will witness in 2023, the biggest unknown is the natural human response. Are we capable of absorbing the onslaught of artificial intelligence, automated cars, and the shift from centralized computing to blockchain-distributed systems?

I predict that we will see an acceleration in the introduction of technology on multiple levels that the social collective is not capable of absorbing.

But the growing concern from the legislative branches of government regarding the self-appointed role of big tech in the definition of "misinformation" could impact their ability to implement some technologies on a broad scale. Such as Artificial Intelligence.

This is new territory, and a new legal framework will be required to address issues such as the ownership of IP rights in AI-generated content.

While a portion of the population will not adopt AI, due to privacy concerns, intellectual property rights will be the rallying point for the pushback against AI. The lack of clear legislation allows the courts to define laws where none exist. And as AI begins to affect the music industry, this will awaken a group that has been visible in the defense of its intellectual property.

The creation and protection of intellectual property should be the focus of every content creator.

The majority of creators are generating income through some form of advertising model. It is time for creators to build a licensing model for their content. Much like the film and music industry is structured where a creator licenses the use of their content by a platform.

Today a creator releases a project and "hopes" it does well. If it does, a shared advertising revenue model is used, with the writer, director, cinematographer, and actor (on-screen talent) , in other words, YOU are getting penny's for your investment. And the social media (hosting) company (YouTube, TikTok, IG, etc.) takes in a huge percentage.

This is not a long-term business model and requires you to take on the burden of financier, creator, and marketing.

Lee Love is a commercial advertising and editorial photographer, artistic director and educator. Although his work includes international magazine publications, television production and photographing everyone from the President of the United States to the CEO of Audi

Corporation, Lee's true passion is teaching photographers to "See Like an Artist." Lee's latest project is the Photo Mentor Academy, where he helps photographers find their creativity, confidence and artistic abilities to bring to the world.

41. Jaime Cohen

Instagram will become an even greater power player amongst the socials.

The algorithm for Reels will become even more refined so new content can be discovered even more easily.

With video becoming more important on the platform, we may see creators start referring to themselves more exclusively as Instagrammers.

Jaime Cohen is an internationally recognized speaker, communication coach and advisor who takes you from good to great for more impactful connection and meaningful communication. Jaime scales your impact from video to beyond virtual. She is also a top LinkedIn content creator and LinkedIn Learning author whose courses have helped close to 100,000 learners.

42. John Kapos

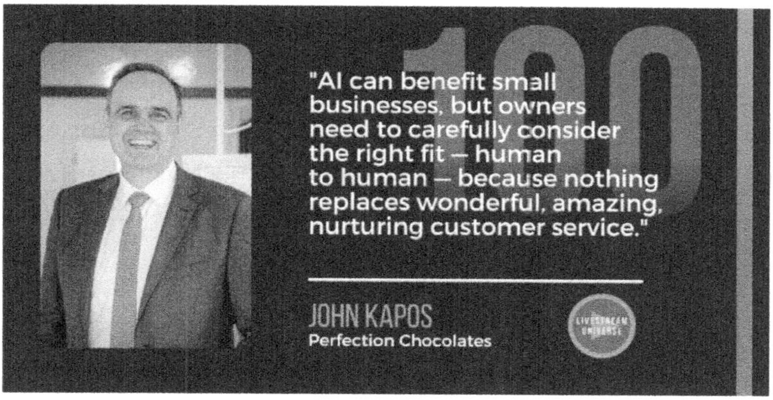

I see AI really emerging in 2023 and beyond. We've seen just recently with Chat GPT and open AI how powerful it is. I've been using it for our small business, Perfection Chocolates, and in my new business in real estate. The benefits have been sensational.

Here some examples of what I think is going to happen with AI in small business:

1) Automation of tasks. AI can help small business automate a variety of tasks, such as data entry, customer service, and marketing. This can free up time for the business owner. I'm pleased to focus on more important tasks such a strategy and innovation.

2) Improve decision-making. AI can analyze large amounts of data and provide insight to recommendations to small businesses. For example, an AI system could help our small retailers optimize their inventory or small manufacturing optimize their production process.

3) Enhanced customer experience. AI could be used to improve customer experience with personalized recommendations in real time support. For example, an AI chat bot could help a small business answer customer questions and provide assistance, 24/7.

4) Increase deficiency. AI can help small businesses, increase efficiency by automating repetitive tasks and providing real-time analytics. For example, an AI system could help a small business identify bottlenecks in its operations and suggest solutions to improve efficiency.

It's important to note that while AI has the potential to benefit small business, it's not a magic solution. Small business owners should carefully consider whether AI is the right fit for their business and how they can best use it.

In a small business, nothing can replace human-to-human, wonderful, amazing, nurturing customer service.

John Kapos a.k.a. Chocolate Johnny is a 3rd generation chocolatier & CEO (Chocolate Eating Officer) with his family business Perfection Chocolates. He helps teach and educate small business owners, corporate companies and entrepreneurs who need to expand, innovate and stand out from the crowd using social media and livestreaming their Voice.

43. Kevin Kolbe

2023 will be the year YouTube establishes itself as the MEDIA HUB for content creators. The place to watch videos—produced and live — listen to podcasts, shop, connect via stories, even host your online course, and more.

No other platform helps creators connect and earn. No other platform will come close to what YouTube will do in 2023.

Kevin Kolbe is a solo content creator and video coach helping creators create, share their gifts and impact the world.

44. Terry Johnson

2023 will be the year of audience-connected content in digital media.

Content creators will discover they have more control over the distribution of their own content. They will gather their audiences on a platform they own and can monetize as they see fit.

They won't have to worry about algorithms suppressing their content. The goal of building an audience that you own and monetize with greater control will grow online.

No more content just for content's sake. Small businesses will be able to understand the link between targeted content and growth. Using AI-powered tools will make it easier for small business owners to connect with their audiences and grow.

Creators will finally discover the power lies in their hands.

Terry Johnson is a marketing strategist and consultant who has a background in advertising, marketing, and technology. She hosts "This

Week's Marketing Chat," a livestream show that covers marketing strategy and social media marketing for small businesses at terryjohnsononline.live.

45. Brian Wallace

I might sound like a broken record when it comes to these yearly predictions, but I'm still bullish when it comes to all things LinkedIn. The true business social network has seriously doubled down on giving proper tools and support when it comes to the creator community — even dedicating actual funds.

Make sure you're willing to invest some time and attention to the platform, and be sure to activate creator mode and gain access to audio rooms, livestreaming, and the especially effective newsletter.

Newsletters have a unique ability to gain reach on and off platform.

Brian Wallace is the Founder and President of NowSourcing, an industry-leading content marketing agency that makes the world's ideas simple, visual, and influential. Brian has been named a Google Small Business Advisor for 2016-present and joined the SXSW Advisory Board in 2019-present.

46. Nicole Sanchez

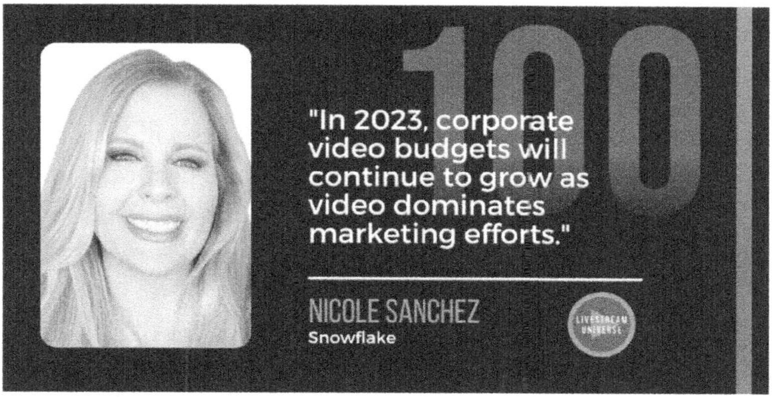

In 2023, corporate video budgets will continue to grow as video dominates marketing efforts. Why? Here are a few statistics:

-Studies show that viewers retain 95% of a video's message compared to 10% when reading text.

Video ads were the number one-way consumers discovered a brand that they later purchased from, and over 90% of people say that they discover new products on YouTube.

-54% of consumers want more video content from brands they support.

Companies will expand to more platforms than ever to reach and relate to their audience. They will especially embrace platforms that use short-form vertical video.

However, companies will still need to produce long-form content.

It is crucial to have this richer content available for audiences that prefer long-form content as well as to have it as a place to point viewers so they can dive deeper into the content they enjoyed in the short-form format.

Making sure to have captions on or available for video content is increasingly essential for accessibility, and over 50% of viewers prefer to have the sound turned off while watching videos.

While many events are in-person again, companies will offer a recorded or livestream option of their events. Intelligent companies delay replays of their event until a few weeks after the in-person event to avoid cannibalizing in-person ticket sales.

Livestreaming will still be one of the best ways that brands can authentically connect with their customers. Livestreams should not be a one-way street. The audience needs to be acknowledged and feel seen.

Influencers of all sizes are still important and can make or break campaigns.

Now go make some great videos!

Nicole Sanchez is Snowflake's senior video producer & YouTube strategist, as well as a beauty and lifestyle YouTube content creator and avid livestreamer.

47. JEFFREY POWERS

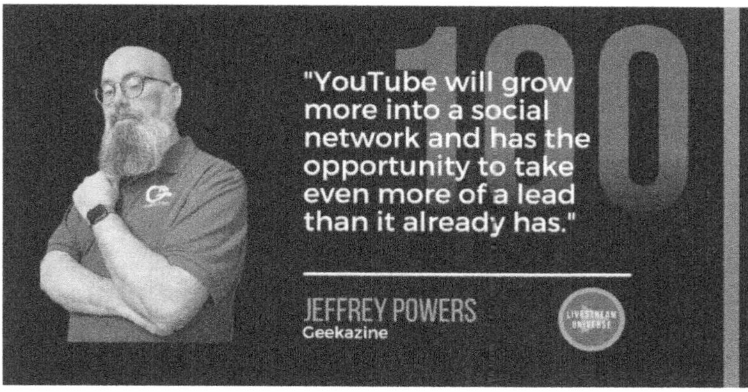

Watch as YouTube will grow more into a social network more than ever before.

While 2022 focused on podcasts and long-form posts, shorts and shopping will be the big focuses this year.

With other social networks pushing boundaries on right and wrong (and what can get posted), YouTube has the opportunity to take even more of a lead than it already has.

I even have to wonder if Alphabet will decide to change the name "YouTube" to something that will encompass the social network change?

Jeffrey Powers is a content creator at Geekazine.com, an Influencer at Amazon Live and creator at YouTube.

48. JENNIFER WATSON

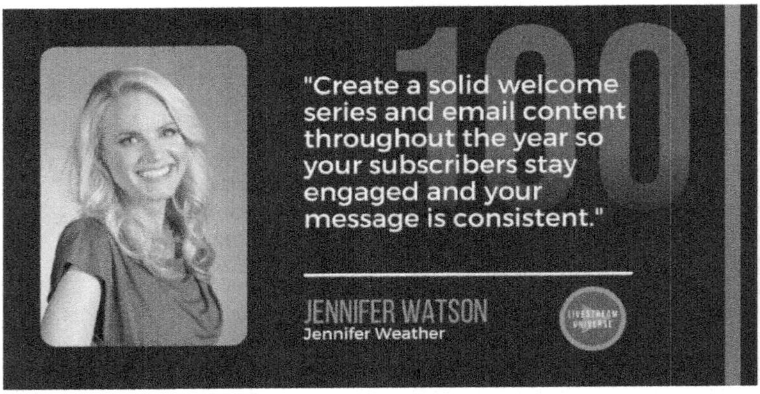

Community building is key, without community, you don't have anything! Social media is a great place to grow a community, but you don't own your social media channels, so that's where email marketing comes in.

You need to be driving the community you're growing on social media to a place you own and one of the best places is an email list. That way, if a social channel shuts down or there is an issue with any of your channels, you still have a way to communicate with your community.

To add to that, you need to have a solid welcome series and consistent email content throughout year, so your subscribers stay engaged and your message is consistent.

Lastly, if you have an online e-commerce store, you NEED to have an abandoned online cart email sequence set up, so if someone gets distracted or forgets something in their online cart, it sends them an email reminding them it's there and to purchase. It's amazing how much potentially lost revenue this can salvage!

Jennifer Watson is a Global Marketing and Social Media Strategist, a Meteorologist on The Weather Channel, Speaker and Podcaster. She motivates audiences from the stage and drives engagement, including launching numerous Facebook Live campaigns for the biggest weather brand, garnering millions of views. She has been featured in Chief Content Marketer Magazine, and on stage at Social Fresh, Social Media Marketing World and Social Shake-Up to name a few.

49. GLENN THE GEEK

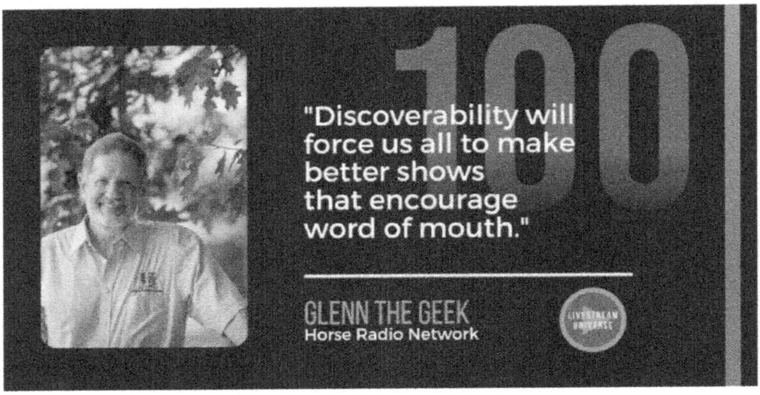

I believe we are going to continue to see more consolidation at the network level of the industry.

As larger companies see value in adding podcasting to their portfolios, they will continue to invest allowing small, medium, and large networks to cash in.

I also think that the growth we have seen in new shows since COVID will continue to decline. At the same time, we will continue to see the quality of new successful shows increase.

Discoverability will force us all to make better shows that encourage word of mouth.

Glenn is the Director of Podcasting at Horse Radio Network, a subsidiary of Equine Network. He is also the host of Horses in the Morning, one of the longest running daily shows in the world at 3100 episodes.

50. Heather Heuman

One of my top predictions for 2023 is attraction marketing through video content. With the ease and simplicity to use the smart phone we all have easy access to, there has never been a better time to leverage more visibility and customer acquisition through video.

Customers are seeking answers to their problems and they're going to YouTube and social media often before they are heading to websites to find answers.

By attracting customers with video, it is especially wise to create helpful video content that offers quick golden nuggets, best practices, and offers ways people can come to learn more if they want.

No longer is it needed to save this video content for a webinar, workshop or week-long live series. I think this is an enormous win for both large brands and equally for small business owners because it can be easily produced and there are so many choices for distribution.

Another top prediction I have for 2023 is the power of real-time human responsiveness. As AI and chatbot bot sequences increase, so are skepticism and a craving by people wanting to be heard, understood, and valued.

By having systems in place to respond promptly in the DMs on social media, as well as to e-mail marketing campaigns, customers will be more loyal to brands and companies that make them feel understood, appreciated, and valued.

As the founder of a boutique social media marketing company, we see first-hand across nearly a dozen brands we work with, that as we respond within the DMs with compassion, helpful information, and attention to the needs of the audience, it shortens the sales cycle and increases the quality of leads for our clients in a variety of industries.

Heather Heuman is a social media strategist, international keynote speaker and author of The Golden Rules of Social Media Marketing, Heather has 23 years of experience in sales, social media and business. Identified as a 2019 Top Business Influencer, Heather has worked with Chick-fil-A as well as hundreds of business CEOs and leaders. She hosts the 'Infinitely More' podcast.

51. Phil Kluba

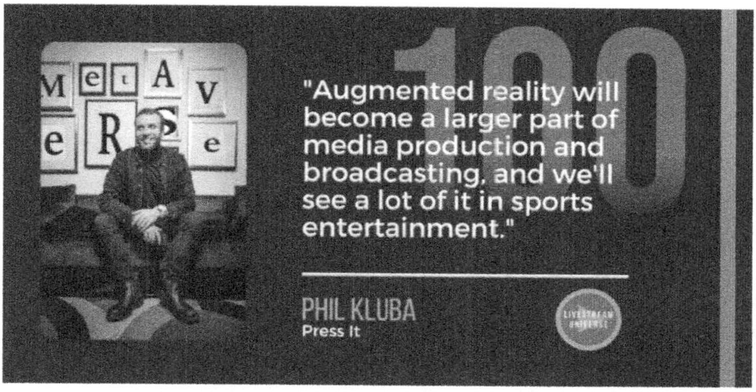

I think we'll see augmented reality become a larger part of media production.

We'll see it on a small scale where independent creators will share filters through social media but, also on a larger scale where broadcasters will implement AR activations.

I think we'll see a lot of it in sports entertainment.

Phil Kluba is the founder & CEO of Press It, a Toronto-based media production company that has spent the past eight years specializing in innovative video, livestreaming and content solutions.

52. MELANIE FALVEY

There is a clear trend of shorter videos, often more creative in style when it comes to shopping or entertaining content. However, long format videos are still the ones that build an audience of buyers, especially for education and service-based businesses.

Using shorts or reels as a bridge towards longer format is a good strategy. User-generated content is being used more and more by brands as people trust other consumers before making their own purchasing decisions.

On Expert Channel TV we will go even bigger on high quality content on OTT, VOD platforms to help clients build their visibility while standing out.

Melanie Falvey is the Founder of Expert Channel TV, an award-winning TV host. She's a TV Host Mentor for entrepreneurs and brands, Video Confidence Coach and award-winning Visibility Strategist.

53. Jim Harold

A significant number of companies will get OUT of podcasting altogether or reduce spending on podcast production as overly optimistic revenue forecasts fall short as the economy softens in 2023. That's the bad news.

The good news is that podcasting as a medium will continue to grow its audience, albeit at a slower pace, and for those who remain dedicated to the medium long-term rewards will follow as the economy inevitably comes out of its slump.

Jim Harold is a professional podcaster at Jim Harold Media LLC, with more than 65 million downloads. He hosts The Paranormal Podcast® and Jim Harold's Campfire, which are among the Top 1% most downloaded podcasts on the industry's largest podcast host, Libsyn. Jim's also the executive producer of Unpleasant Dreams and the author of five volumes of "Jim Harold's Campfire: True Ghost Stories" (All Kindle #1 Supernatural Best Sellers).

54. Ian Anderson Gray

In the past year, we have seen a significant increase in new features for livestreaming tools. We can expect to see even more new features, particularly for repurposing content such as local recording, isolated audio and video tracks, guest booking, and podcast host integration.

There will also be a focus on making repurposing easier through transcription-based editing.

In 2023, AI will be a major focus for live streaming tools. We can expect to see tools that help us plan and generate ideas for our live shows, as well as tools that clean up transcriptions and create summaries or show notes.

AI will also be used to improve the quality of audio and video, create graphics, and assist with branding. While there may be a trend towards using AI-generated avatars as hosts, there will also be a push for authentic content.

Overall, we can expect AI to play a significant role in improving the livestreaming experience both on and off screen.

Ian Anderson Gray is the founder of the Confident Live Marketing Academy and host of the Confident Live Marketing Podcast. He's an international speaker, trainer, teacher and consultant, having spoken at conferences such as Social Media Marketing World, Content Marketing World, VidFest and more.

55. Melody Johnson

My prediction for 2023 is that more consumers will rely on online shopping to fulfill their needs.

They will look for convenience and rapid service.

Melody Johnson is an influencer and online coach to over 2K creators. She has not only reached A-List influencer on the Amazon Live platform, but has grown her audience on stages, where she's spoken about the importance of becoming your own brand and how to market yourself as an influencer.

56. MARK GAWITH

2023 is going to continue the trend to democratize video content.

Video will continue to be the driving force behind social content and marketing efforts at companies, as well as the ways brands create authority, authenticity and build reputations.

The development of technologies to create more video content will start to come into focus with AI being a large driver of this. Taking long-form content and turning it into short-form content with a few mouse clicks will be the norm.

Doing more with the same, or more with less, will be the mantra for video companies as each one tries to stay ahead of the other with massive amounts of development thrown at this.

One or two leaders will emerge but there will be hundreds of "imitators."

Live video will continue to play a part, but takes a backseat to more TikTok style and YouTube Shorts style content.

Marc Gawith has spent the last 7 years in the video space with many of those years focused on building companies that help users create a "better" livestream. He has worked with all of the influential platforms, key influencers and first movers in the space. He is also a bit of a creator in his own right as well as a "dabbling" live video producer.

57. Professor Nez

Short-form vertical video will be the primary focus of all the major platforms in the creator economy in 2023. And the ad spend behind it will be record-breaking.

Creators, businesses, and brands that don't leverage short-form video in 2023 will be leaving a lot of free traffic and money on the table.

Professor Nez is a full-time content creator helping creators, thought leaders and brands monetize in the creator economy. He was selected as one of the top creators by LinkedIn for the inaugural LinkedIn Creator Accelerator Program (#first50) for the Creator Economy.

58. Bridgetti Lim Banda

In a world with growing distrust of mainstream media, more and more people will rely on livestream content creators to become the source of information that is not skewed by hidden agendas or propaganda.

Not only will livestreaming become a trusted source of information, but it will also become the preferred source of entertainment and education.

It is therefore more important than ever to embrace and promote responsible livestreaming and for collaboration to trump competition, to be yourself and attract people to your unique personality.

Bridgetti Lim Banda is an Executive Livestream Producer, Livestream and Podcast Host, Global Goodwill Ambassador, international best-selling Author, and an Invisible Disabilities Advocate. She helps people level up and scale their impact so they look, sound and introduce themselves better on camera so they can build trust and authority and increase brand recognition. Bridgetti's work with the Cape Town Water Crisis has been featured in print as well as on international radio and television.

59. Eddie Garrison

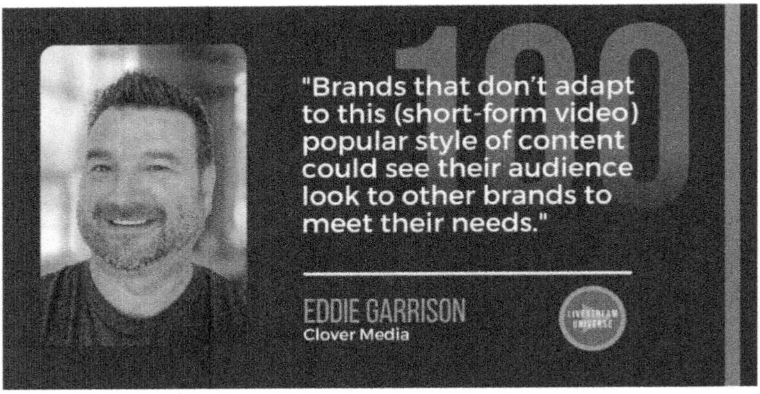

The rise of short-form video content can't be denied.

Brands that don't adapt to this popular style of content could see their audience look to other brands for their needs.

Utilizing this new style of content in conjunction with lead generation will help businesses capture their audiences contact information to communicate with them directly through their targeted, personalized messaging.

Eddie Garrison is a Social Media Strategist, Award-Winning Video Creator & 3-time author helping brands grow their digital presence with real world strategy, design and results.

60. Brad Friedman

In 2023 you'll find me in the Metaverse. I'm predicting we'll start to see more and more creators experimenting and ultimately taking advantage of the Metaverse.

I will be meeting clients in the Metaverse conference room I've built. I will be livestreaming, podcasting, holding virtual events, and recording video in the Metaverse. And I'm predicting I won't be the only one.

There is discontent with many of our current "go-to" digital platforms. Platforms that we don't have control over. I believe we are going to see people looking for new places to build community and create. One of those places is the Metaverse.

Whether it is with the use of a headset like the Meta Quest 2 or from your desktop, the Metaverse is very accessible. And, in the Metaverse, we can control our own spaces and build our own communities.

Digital transactions are taking place in these digital worlds right now.

We are seeing a rise in the use of NFTs and Smart Contracts to transact business. More and more creators of all types are going to be taking advantage of Web3 and the Metaverse in the next several years. And many people are going to be leveraging Web3 and the Metaverse starting in 2023.

I believe you'd be smart to stake your claim now!

Brad Friedman is a digital marketing strategist, coach, author, podcaster and speaker. Brad works with professional services providers and business owners to help them generate revenue in traditional online venues and in the Metaverse.

61. Lee Uehara

AI will feature more prominently when creating content this year.

Content creators will use more Artificial Intelligence (AI), to generate items for their shows and other projects.

For example, with the new app ChatGPT, entire blogs, posts, and eBooks can be created with just a few key words.

With Lensa iOS regular photographs are morphed into glamorous headshots.

However, there are caveats with any AI app. Some verification of results is necessary.

With ChatGPT, inaccurate information could be included in an essay, answer or command.

For other AI apps, original files may be used to further refine the software. This could lead to a question of ownership, among other things.

Lee Uehara, serial podcaster and host of House of Lee NYC, is the co-founder of: AAP (Asian American Podcasters Association); the Golden Crane Podcast Awards; Podthon.com; and NYC Podcasters.com, an indie collective. Lee is also a rockstar photographer to podcasters and speakers.

62. JOHN PRETTO

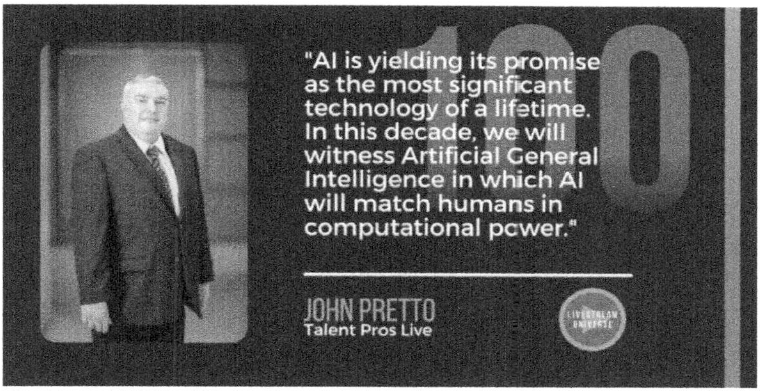

After five decades of work, Artificial Intelligence (AI) is yielding its promise as the most significant technology of a lifetime. Several groundbreaking applications from OpenAI and DeepMind have gained huge usage in 2022 and in this decade, we will witness Artificial General Intelligence in which AI will match humans in computational power.

AI has already solved checkers, chess, go and many other games that people thought "unsolvable," and this is just the beginning.

AI will be helpful in curing disease, creating new super materials, solving complex math problems, and much, much more.

Be careful, however, because there will be a "bubble" of AI startups and a crash.

Additionally, AI will dominate creative fields such as composition, illustration, music and movie making.

These endeavors will surpass humans this decade.

Conversely, we will see the rise of the "neo luddites" to oppose such technology, and society will become bifurcated over the use of AI.

Society will try and hinder AI to thwart any effort of AI getting out of control. More on this next year.

John Pretto is known as the "Grandfather" of streaming, starting a company with proprietary streaming technology in 1996. John's career spans three decades and many technology launches.

63. Beth Granger

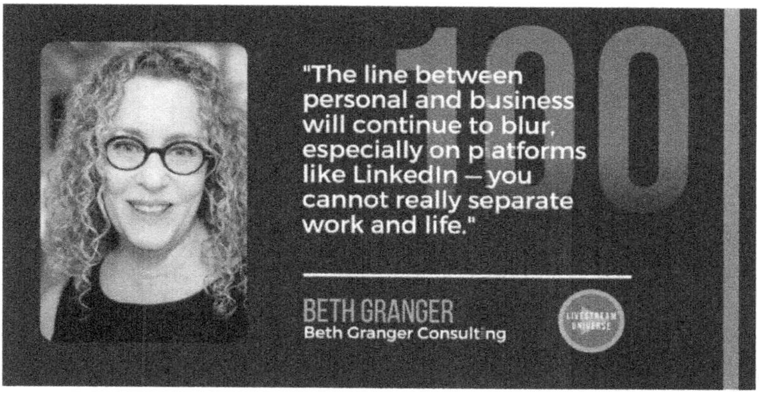

I predict that the line between personal and business will continue to blur, especially on platforms like LinkedIn.

If we learned anything during the pandemic, it was that you cannot really separate work and life.

I predict that new tools or new features in tools we already use will make it easier to create content, such as AI content development for your first draft.

Beth Granger is a trainer, consultant and speaker who works with organizations and individuals who want to grow their business and brand by unleashing the power of LinkedIn and networking. Beth is also an Exactly What to Say™ Certified Guide, bringing her clients magic words for influence and impact during critical conversations.

64. Gord Isman

The applications of AI technology for video editing are evolving very fast, putting more powerful and sophisticated features into the hands of creators.

What factors will help amplify the creator's interest to dive in, adapt and reap the benefits?

Time-saving benefits are a primary factor in adoption. For example, features like audio sync, audio cleanup, color matching, auto color grading, face detection, audio ducking for music, auto reframing in motion graphics — to name a few — save time, not to mention the tedious aspects of some of these tasks if done manually.

The ease of use is also essential because some people have shorter attention spans than others and different skill levels. So, if there's a steep learning curve for a particular AI feature, one may decide to pass on it.

The ability to produce higher-quality results with minimal input from the individual is very appealing, especially if one can create a more professional finished video. For example, features like audio cleanup, object/background removal, or audio ducking with music require very little input to improve the quality of the finished video.

Other factors that may influence the adoption of AI video editing tools include:

The cost of the tools.

The level of support and resources available.

The ability of the tools to integrate seamlessly with other software and hardware.

Seamless integration is something I always hope for. I don't want my workflow to require me to take the output from one tool into another and then back again to finish my video. Currently, depending on the project, my workflow may need me to use several different tools to get the finished product I desire.

Some video editors may be concerned about the accuracy of AI tools, but as these tools continue to evolve and improve, they are likely to become more reliable and trustworthy. For example, many video editing tools now include the feature to cut out filler words and long pauses from audio/video media. If applying this feature creates more of a mess to clean up due to "what happens," one might feel that there's a lack of accuracy, so the net result is not worthwhile. The good news is that AI algorithms and learning keep evolving and improving accuracy, quality and usability.

Others may be concerned about losing control over the creative process, but it is important to remember that AI tools are designed to assist and augment the work of human editors, rather than replace them.

We should embrace these technologies as they are maturing nicely. If we don't dive, adopt, and adapt, we're likely to get left behind and be labelled as video content creators that become part of the mediocre quality crowd, producing a lesser quality product, and with less efficiency in our workflow.

Education and experience can help to overcome any fears or hesitancies that video editors may have about using AI video editing tools.

Video editors can also start with simple, easy-to-use AI tools and gradually work their way up to more advanced tools as they become more comfortable with the technology.

By keeping an open mind and being willing to experiment with different AI tools and techniques, video editors can discover the benefits of using AI in their workflow and overcome any fears or hesitancies that they may have.

And let's remember that as a content creator, there are benefits to utilizing several other AI solutions (e.g. generative AI) that can help to raise the bar of the quality of our results with greater efficiency. For example, ChatGPT, and various copywriting AI solutions based on GPT3 (and forthcoming GPT4), help us greatly in areas like content research, script writing, and copywriting.

I'm all in on the AI journey!

Gord Isman is a Video Strategist, Video Editor and YouTube Consultant. He helps entrepreneurs, coaches and small businesses utilize video to get their message out to the world.

65. JUNAID AHMED

The year 2023 is the year of the podcast. As more and more businesses and entrepreneurs are creating communities, podcasting is the fastest way to build authority and connect deeply with our audience.

YouTube is offering up to $50k to podcasts and up to $300k to podcast networks to enable them to bring in video podcasts.

Podcasts and YouTube will continue to bring people interested in new topics together.

Focus on brand and community podcasts because they have the longest retention of audience when compared to social media, short videos and image posts.

Podcasting is also one of the easiest ways to create long-form organic pillar content for all forms of content to build off of.

Plus, compared to writing blogposts, we are capable of speaking at 300-400 words per minute as compared to writing at a maximum of 120 words per minute.

If you're looking to build authority, start with your podcast and build a rapport with the industry on your expertise.

After bringing over 400 guests on the Hacks & Hobbies podcast, I find it's been an experience which cannot be learned overnight.

Start now with your podcast and build your authority and authentic connections with your audience.

Junaid Ahmed is a Home Studio Architect, building studios designed to capture high quality video for podcast hosts, content creators and YouTubers.

66. MARIO FACHINI

Polish or perish. The core of what you and I know and love is nothing new.

E-Comm has been around just as long as the internet. I've been investing in training since 2002. Video's been here pretty much since I was born— not gonna date myself— but it is decades old, fair to say?

Granted I started EAInterviews in 2019 and podcasting was nothing new then. Honestly, it felt "late to the game" but compared to what?

You may feel that way now, especially if you're just getting started. Congrats and welcome! Maybe you're considering starting a YouTube channel? Or (I say very lovingly with all the affection as I can muster in text) perhaps you're crazy enough to livestream? You could be required to do both as components of your next virtual event.

Don't play small and don't skimp. It's never, "Here's all you need to do."

Think about your ideal reader/listener/watcher and do what you know is right in your heart for them, even if you feel like you're the only one doing it.

2023 is the year you will need to be polished – or your efforts will perish.

Perhaps "everything seems like it's working," but I heard long ago and it still holds true, "You can fool some of the people some of the time, but you can't fool all of the people all of the time."

I could and have written books in very descriptive detail expanding on the premise of this thought. And if you want to argue the point or debate it, I'm open. However, let me ask you one thing first, even if you could win the argument for why you shouldn't do your best for your business and audience. Is that really something you'd be proud of striving for?

Mario Fachini is a 2x #1 International Bestselling Author, Speaker, and Top 100 Entrepreneurship Host of EAInterviews.com. Fachini reveals insights from over a decade's worth of book publishing strategy, public speaking, livestreaming, and video production experienced from helping hundreds of authors with their launches, plus five of his own books, shows, and training. Fachini takes you on the journey of not only what it takes, but also what doesn't matter at all on your journey, so you can focus on what does, your message, making a bigger impact—and profits, while serving your audience.

67. Mike Gingerich

The evolution of online shopping will continue with advancements in 2023.

You will see more and more livestreams for selling, and more video with one touch purchasing.

Currently, TikTok is leading in this, but Instagram is trying to keep pace. An interesting development will be the growth of a "TikTokesque" feed on Amazon as the mighty e-commerce giant moves into social selling and social media even further!

Traditional e-commerce websites must shift and increase one-touch purchasing with tools like Apple Pay, Shopify, and Google Pay.

The key to repeat success will be a seamless and easy experience that extends through shipping and delivery.

This means out with antiquated, slow shipping with USPS and stores providing a better, complete experience with shipping via UPS and FedEx.

In essence, in 2023, the entire process of online shopping from video to doorstep delivery for the highest-rated stores will become more interactive, seamless and painless.

Mike Gingerich runs 3 professional blogs on business, lifestyle and health while participating in triathlons and half marathons to raise money for worthy global causes.

68. Kendra Ramirez

I am most excited about Open AI's ChatGPT. It is a game changer. My prediction is the continued evolution of AI research, analysis, and deployment for our industry and the B2B industries that we serve.

AI isn't replacing us. It is elevating our work. Much like watching a construction worker operating a Bobcat, it is about efficiencies and human and machine connection. Embrace change and stay curious.

Kendra Ramirez is globally recognized on the home page of LinkedIn.com, Women of Influence Award Winner, John Barrett Entrepreneur Vision Award recipient, Cincy Chic Woman of the Year, AMA Marketing Legend, best-selling author of Marketing Fusion and Micro Shift books and a finalist for the Social Media Innovator of the Year. Since 2005, she has helped hundreds of organizations successfully leverage digital technologies. Kendra has spent more than 15 years in technology companies as a change agent in sales, marketing and recruiting leadership for B2B customers.

69. Mike Wilkerson

One of the most rewarding elements of working in podcasting as long as I have is to see how it evolves.

The most recent presentation that continues to fold-in "LIVE" elements to varying degrees of success, is the "LIVE" element to an existing, produced podcast/video program.

While you have the "ongoing/produced effort" that gives and satisfies new and old listeners, you're also seeing MANY programs offer a "LIVE" element. By adding this element, typically a far more informal sit-down with one or more of the program hosts to just chat to talk about their lives, the things they see around them, and to start additional conversations where listeners are invited in.

This extra frosting offers something extra sweet, like a wonderfully crafted chocolate cake.

Raise your hand if more frosting somehow hurts a bite of cake.

If you haven't considered this, it's time to do it and enjoy as you educate/baste new listener experiences and enhance the flavor of existing listeners.

This is another opportunity to share more of the "behind the curtain" that makes sharing perspectives so satisfying on multiple levels.

Mike Wilkerson is an OG Podcaster in St. Louis area who has crafted, hosted, produced and promoted podcasts of every kind since 2005. His growing Podcast-service-based empire includes a voiceover academy and service (The Voice Farm), a podcast editing academy and service (The Editor Corps) and a program promotion engine (The Podcaster Matrix). He has created 4 retail-based Podcast Studios to capture content to propel each vein of his business. For a fifth studio, a unique and mobile "Podcast Bug"— a 1974 custom Super Beetle with a recording studio built into the front of it— is at the ready.

70. Doug Cohen

I believe that no matter what advances take place in e-commerce trends, livestreaming, podcasting, Metaverse etc., the main social platforms where people share this content always continues to be central to how people connect and build online.

So, for 2023 I'll focus on the prediction that things are about to get interesting again with a new round of contenders/pretenders vying to be a "next big thing."

I think a new one will gain some real traction with the volatility around Twitter, in particular, and the landscape will shift a bit yet again.

This isn't EARTH SHATTERING DISRUPTION that I'm predicting, but whether it's Post, Mastodon, Hive, or some other platform, I do think one will emerge as a player.

Doug Cohen is a Metro Detroit-based entrepreneur with 3 different businesses: A brick & mortar, a digital marketing brand, and a fun e-commerce biz he just launched. A "relatively well-known livestreamer"

😉 going back to the OG Periscope days, he has been featured as a guest on many livestreams, podcasts and publications. He is a former Michigan Wolverine OLB, a history buff, and the lead singer of Vintage Playboy. He is married to Ally & they have two awesome grown kids.

71. Dr. Aikyna Finch

Livestream will evolve to include more VR.

There will be more VR conferences and podcasts as well.

There will also be hybrid conferences as people still want the option to not travel.

LinkedIn Audio live use will pick up. LinkedIn will add even more features to make it more engaging overall.

Dr. Aikyna Finch is an internationally certified and award-winning Life and Social Media coach with a background in Business, DEIB, Education, Marketing and Technology.

72. Chris Curran

Business and career opportunities for remote podcast producers/engineers will skyrocket!

Podcast producers who are professionally trained will have even more ways to earn six figures from the comfort of their homes.

Chris Curran is an audio engineer, podcast producer, founder and lead instructor at Podcast Engineering School.

73. Tim Sohn

During the pandemic, you saw people sharing more personal aspects of their lives on social media. In 2023, you will see this happen even more.

In terms of brands, you will see more brands using less polished video and more organic video that everyday folks can relate to. It's all about relatability.

Tim Sohn is a husband, dad, lover of Dunkin' iced coffee and cancer survivor. He is also the founder and CEO of Sohn Social Media Solutions, where he is the powerful man behind the curtain producing livestreamed shows and virtual events so you can just show up and have amazing conversations and don't have to worry about the tech or behind the scenes.

74. AMY WOODS

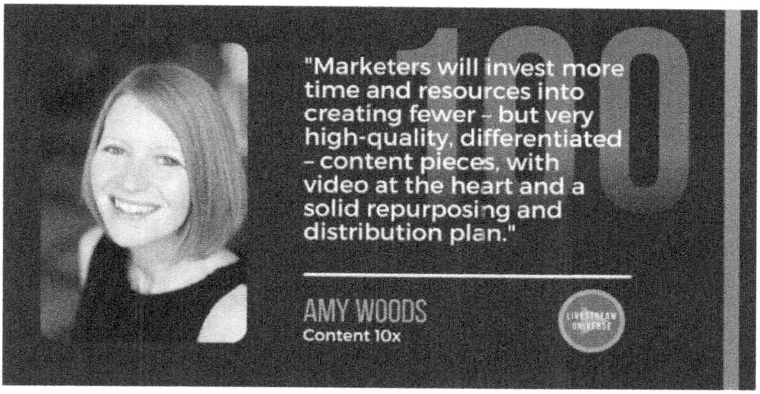

There will be a growing realization that "more, more, more" isn't the answer. The focus should be on doing more with less.

Marketers will invest more time and resources into creating fewer — but very high-quality, differentiated — content pieces, with video at the heart and a solid repurposing and distribution plan.

Amy Woods is an expert in content repurposing and the CEO and Founder of Content 10x – a niche creative agency and the first content repurposing agency in the world. Amy is the host of The Content 10x Podcast and B2B Content Strategist, the author of best-selling book "Content 10x: More Content, Less Time, Maximum Results" and an international speaker.

75. Kyle M. Bondo

Podcasting 2.0 is an industry disruption concept heralded by Adam Curry and Dave Jones via their podcast, Podcasting 2.0, along with a dozen or so developers, engineers, and evangelists — is summed up by a quote from Tom Rossi, "[100-percent focus] on improving the podcasting experience in an open-standard way that allows every player to innovate faster and better than any one company could do on their own. This is our best bet at avoiding one company emerging as the monopoly of podcasting."

Player innovation — that's a big deal, mainly because it's the pointy end of the podcasting spear that directly impacts the listener. There is no point to producing a podcast without a way for your listener to find and play that podcast.

Why is Podcasting 2.0 so important? Because it is more than a technology change. It is a philosophical change to how we use and understand podcasting. It is imagining a world where big tech no longer sets the rules for podcasting, a real move back to open-source podcasting.

The adoption of the Podcasting 2.0 philosophy has already started. Podcasters, podcast media hosts and podcast app developers have been building, failing, building, failing, and building again for years. However, within the past two years, they are beginning to see the innovation they have been working for. After finding 10,000 ways to not innovate podcasting, they found the one way to do it: The Podcasting 2.0 namespace.

For the longest time, the Apple's iTunes namespace has been driving the podcasting train. Well, Apple and their 500-pound gorilla called the iPhone. Within the Apple iPhone closed ecosystem, Apple has controlled the main way podcast media is read into podcast players by greatly influencing what data will and will not be used by Apple podcast players.

This is called gatekeeping and Apple can gatekeep with the best of them. Their market power gives them the capability (and the leverage) to push podcasters, podcast media companies, and podcast player developers into a direction that benefits only Apple.

However, for far too many years, Apple has done something far worse then use their influence to control podcasting: they have done nothing.

Nada.

Zilch.

They forced the podcasting industrial complex to adopt a format that benefited Apple apps and then quietly turned their podcasting attention on autopilot.

Sound familiar?

Since then, that format — that iTunes namespace — has atrophied. We all waited. Waited for the sleeping 500-pound gorilla to wake up and realize that changes were needed. But nothing happened. We asked, we demanded, we even begged. For over a decade, we accepted the tyranny of big tech and got nothing. And the reason we got nothing was simple: Apple doesn't have to do anything. They've optimized the market to their benefit.

Steve Jobs even told the world that podcasting was, "... the concatenation of the word iPod and Broadcasting" at the All Things D3 conference in 2005, he, along with Apple, thought they owned this technology too.

However, the adoption of Podcasting 2.0 namespace will mean the end of that control because it will represent more than just Apple. It will represent the entire podcasting eco-system that has grown despite Apple's reluctance to innovate.

I believe that when the podcasting namespace blows wide open in 2023, three very amazing things will happen:

1. It will mark the end of gatekeepers. It will take central control away from big tech and remove the constraints over what a podcast feed can deliver to the listener via a Podcasting 2.0 enabled podcast player. In other words, it will allow new relationships to form between the podcaster creator and their listeners without a tyrannical middleman dictating how that connection will be made.

2. It will allow podcasting application developers to experiment with the overall listener experience by leveraging the influx of new podcasting 2.0 information. The current Podcasting 2.0 namespace already includes tags that allow for new listening behaviors. These behaviors include "value for value" connections where listeners can provide direct messages and monetary feedback to podcasters.

3. Podcasting will start growing again. Experiments in just what the Podcasting 2.0 namespace can do will open podcasting to new listener experiences that we have yet to think up. The namespace has already started expanding into monetization/funding via crypto currency, geo-location for local podcast directories, alternate version delivery, live podcasting notifications, transcript and chapter augmentation, and even expansive biographies of hosts and guests. Or, to sum it up in a word: innovation!

One simple change, the adoption of Podcasting 2.0 namespace, will be the cornerstone on which the next podcasting revolution will be built.

This is the namespace space race.

On one side we have the Podcasting 2.0 namespace, out in the open for everyone to use, versus the closed ecosystem of the big tech companies that make up the podcasting industrial complex.

I strongly believe the future of podcasting depends on open, community-driven development of the Podcasting 2.0 namespace. As more podcasters learn about this movement, it will create more and more healthy debate on what the future of podcasting should look like.

So, I leave you with this final thought: who controls podcasting?

I see podcasting as the last gasp of true intellectual freedom over the next year. Those who love podcasting will have to ultimately decide where podcasting innovation lives.

I once believed that big tech would be the ones to innovate podcasting. Unfortunately, I have come to realize that these companies have become the same bloated big companies they once claimed to be revolting against. They are now the new gatekeepers who focus only on optimizing the status quo or cancelling those that displease them.

No, the real innovation in podcasting will not come from big tech. I see now that true podcasting innovation will have to come from the podcasting community itself. It will be the rag-tag bunch of podcasting enthusiasts who have been tinkering in their internet garages, experimenting, testing, and learning through trial and error, what the future of podcasting can be.

True podcasting innovation will be driven by those that stuck with it until they had something that was worth showing to the world.

This is my prediction for podcasting: An explosion in the adoption of the Podcasting 2.0 namespace.

Let the namespace space race begin.

Kyle Bondo is dedicated to simplifying the complicated world of Podcasting. As the Chief Creative at Gagglepod and the founder of Oncetold, Kyle has been podcasting for 7 years, designed and produced several podcasts, and hopes to make a small dent in your podcasting universe by evangelizing the principles behind good podcast design.

76. Chris Strub

As consumers get wise about their spending in the face of macroeconomic challenges, the increasingly vast landscape of streaming services will continue to consolidate.

Yes, only the strong will survive — but strength may end up being more about acquisitional prowess than content quality.

Independent creators will struggle to keep pace as these professional streaming conglomerates identify clever ways to pump orchestrated, edited content into traditionally organic channels, strengthening their grip on the attention of the public at large.

Chris Strub is the Giving Day Guy, working with dozens of fundraising events around the USA, helping raise hundreds of millions of dollars for tens of thousands of nonprofits in various communities through livestreaming video.

77. Sue-Ann Bubacz

The digital dance is in full swing making music to suit any taste and style.

For content creators, this means unlimited opportunities to connect with audiences easily and uniquely. So, grab your dancing shoes and get moving!

Whether you focus on written, video, visual, or audio content, standing out continues to be more difficult yet imperative. Fortunately, leveraging platforms, software, and tools to enhance your digital performance are readily available to use and optimize. Go ahead and take the stage.

While you need to know the exact steps for the foxtrot, salsa, or ballroom, freestyling with creativity and innovation is where you really wow the crowds. Don't be afraid to experiment!

2023 offers a turbulent yet exciting chance to test, rethink, and try things, especially digitally. Don't let stage fright stop you.

But the real issue for establishing a credible digital presence is about how you connect the dots to reach people and create business. You can't do the tango alone.

Statistics support a written strategy to seal the deal to go for the gold, even if this "new" advice is tried and true good business. Ask yourself, "Are you dancing for fun, competitively, or to star in the ballet?"

Because, for success, a strategic approach, from tech to touch, is a key business driver in a furiously evolving digital landscape.

This is the think-before-you-fly part, and there's never been a better time to regroup and reinforce your core mission and plan, a key to reaching and exceeding goals.

Still, the hot ticket is taking an omnichannel (no matter the channel, old or new) approach using customized mixed mediums, social digital tactics, and technology in tandem. Think selectively and strive to excel on your key platforms.

It's a good idea, especially with the instability and changes in social platforms, to bring all channel traffic back to your website for closer nurturing, direct contact, interaction, and eventually conversion.

No matter how digital or meta you are, bring your community home as a priority in 2023 more than ever before! It looks like it's time to accept modern dance and mixed media artforms and performances, sometimes a rather radical disruption.

The bottom line is no matter how advanced your Marstack, AI, Web3, cool gear, costumes, or other technologies are, your ability to communicate with real people matters most. Perhaps a case of something old is new again?

Think creatively about how your marketing techniques can leverage marketing motivation, the ability to get people excited to take action. Applause, applause, please.

Five stars if your digital dance feels immersive, tugs on emotions, tells a story, and otherwise draws people in!

Creating a loyal community with your mesmerizing moves and building relationships through your digital strategy delivers clients and business growth.

But, *as always*, an emphasis on connecting with humans is the ultimate goal. H2H — the mirrored ball trophy — for the win!

P.S. Go ahead and struggle with change, or you can enjoy the dance while learning and evolving. Best in 2023...

Sue-Ann Bubacz is a digital content strategist and writer/editor who works with best-selling authors, media personalities and prominent artists to enhance their communication, elevate their influence and transform their creative vision into tangible business results.

78. Andrew Kavanagh

I predict in 2023 that you will see a boom of all things AI - Artificial Intelligence.

I love that it is being implemented in Adobe apps and improving almost daily. It helps those with disabilities, will improve audio, video and writing in various apps, and for social networks.

Yet I am concerned about the fact that it will or already has taken away jobs from some very skilled illustrators, steals the style of established artists with no credit to those artists, has thrown copyright concerns to the wind, and my biggest concern of all is how AI will be used by scammers to scam people via email, phone, or a deeper ability to hack into people's banks and someday even into areas of the government or agencies of intelligence and private information.

In 2023 I think there will also be many legal cases regarding copyright and usage of AI. Get ready - cause it's gonna be a wild ride!

Andrew Kavanagh is a Digital Artist and Photoshop Tutor. He does Photo Compositing and Photo Retouching in Los Angeles, CA. He is also the creator of various Photoshop and Digital Art focused Facebook groups. Kavanagh's Photoshop and Lightroom Group has more than 320,000 Members.

79. Brian Schulman

The 2023 headline continues to be inflation. Rising costs are causing worries and heartache for many individuals, families and wallets.

During tough times, it's the smallest things that make the biggest impact. Simple acts like a smile, a laugh, a 'hello,' have the power to provide a sense of happiness and normalcy.

Things that transport us from our reality, taking our minds and hearts away from stress and worry, are invaluable. Click a button, a stream, find a game where athletes give us hope, a community, and someone to root for. Movies that take us to another world, drawing us in with excitement to get us cheering together for the 'good guys.'

Concerts, where our hearts beat together to the rhythm drowning out the negative voices in our heads. Stand-up comedy shows, where the realities of the world escape our minds and the pains in our sides are caused not by the world we live in, but by the laughter it elicits.

Today, with digital media, opportunities for escape and community building are in the creator's hands. Consider LIVE shows, short videos and series' where hearts smile because community gathers.

A digital media content strategy that creates chances to support, uplift, share life with one another and feel less alone.

In 2023, and years to come, providing opportunities for happiness, moments of forgetting your worries, and enjoying a sense of belonging whether through LIVE shows, native video or other digital media outlets will be the key to voicing your vibe and showing up in a way that brings positivity where it may not otherwise be felt.

While the economy is being challenged, online influence has become a necessity. As humans with a heartbeat, creators and entrepreneurs, we can provide a break from reality. Gifting people much needed time for feeling joy and hope, smiling and escaping, are ways that help us all thrive during tough times.

While making a greater impact you will see the fruits of your labor like never before.

Brian Schulman is a 7-time #1 Best-Selling Author, Voice Your Vibe Founder & CEO, internationally-known Keynote Speaker, 3-time Top 50 Most Impactful People of Linkedin (out of 1 billion), and Linkedin Top Voice. He is known as the Godfather & Pioneer of LinkedIn Video and one of the world's premier livestreaming and video marketing experts. His insights & 2 Global Award-Winning Live shows have been seen in 120+ countries & featured on NASDAQ, Forbes, ThriveGlobal, YahooFinance, Viacom, ROKUTV, AmazonFire, TheCW, multiple #1 best-selling books, syndicated on a SmartTV Network and hundreds of shows & podcasts, reaching millions world-wide.

80. Katie Hornor

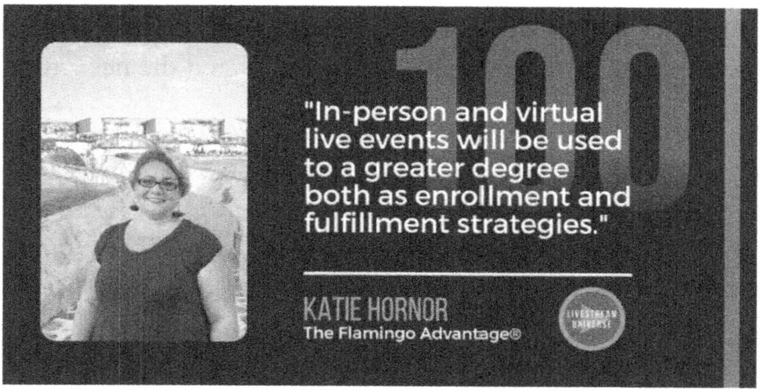

I believe that coaches, consultants and service providers in 2023 will be utilizing in-person and virtual live events to a greater degree both as enrollment and fulfillment strategies.

Virtual events allow organizations to reach a larger and more diverse audience, as attendees can participate from anywhere in the world. They also offer unique benefits for attendees, such as not having the expense or time constraints of travel and the ability to still engage in real time with the speakers and other attendees.

At the same time, in-person events will likely continue to play a significant role in the world of business and marketing, as they offer a unique and immersive face to face experience that simply cannot be fully replicated digitally.

I also believe that as more and more AI technology is introduced to the marketplace, the necessity and value of human interactions, accountability, and wisdom, prior to and after the sale, will increase in both exclusivity and value.

Every business needs to be asking how to be more human in our client experience practices even as we embrace the speed and ease the AI advances provide.

In the end, we still do business with humans, and humans want to feel seen, heard and loved even when being offered the next step.

Through her Flamingo Advantage® Framework for client experience & marketing, Katie Hornor has helped over 8,000 coaches, consultants and service providers learn to maximize the impact of their expertise by scaling their online programs without compromising their Christian faith, their core values or their non-negotiables.

81. BILLY THORPE

The days of needing thousands or hundreds of thousands of insert metric: (downloads, views, followers, etc.) to make a significant amount of money as a content creator are over.

With the continued growth of platforms like Amazon Live and Walmart's Creator platform, content creators of all levels will be able to create multiple streams of revenue quickly.

It's important to mention this will only happen for those who consistently apply the elementary things: showing up, doing the work, being a student and producing more than they consume.

Billy Thorpe is an Amazon Live Influencer, Podcast Host and Monetization Strategist for content creators.

82. Julia Jornsay-Silverberg

I believe that 2023 is going to be the year of the micro influencer; we are going to see so many more brands lean into the power of influencer marketing and start working directly with smaller creators as a way of expanding the reach and awareness for their brand.

I also think with this shift will come an emphasis on vulnerable, authentic content.

People are tired of seeing overly produced and edited photos and videos; people want to see real, relatable people, doing real things in real time, and I think we'll see a ton of smaller creators gain much larger audiences as a result.

Julia Jornsay-Silverberg is a social media enthusiast who loves to excite, inspire, and motivate people to cut through the digital clutter by being authentic and vulnerable online. She has a diverse background working for The Buffalo News, the Buffalo Niagara Partnership and Telesco Creative Group. As the 2018 recipient of the "Emerging Alumni Award," granted by the University at Buffalo School of Management

Alumni Association, Julia has more than 10 years of experience working in digital marketing and is known for delivering powerful presentations that leave people feeling excited and empowered.

83. Christian Karasiewicz

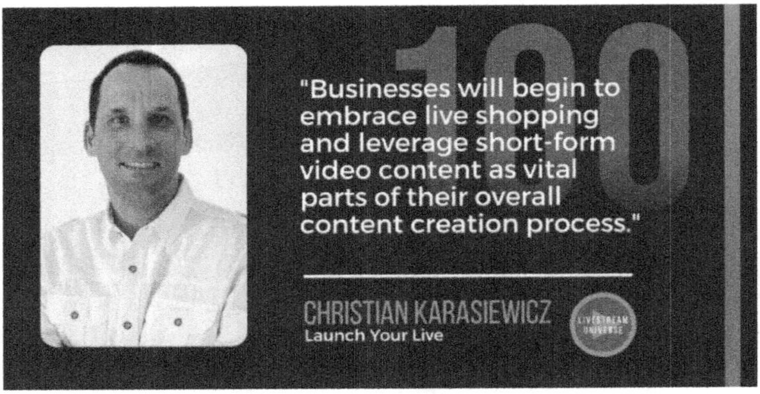

Businesses will begin to embrace live shopping and leverage short-form video content as vital parts of their overall content creation process.

As video content has grown in importance, we'll see more companies improve their efficiency and repurpose their written content into pre-recorded videos, podcasts, and livestreams as effective means to reach their target customer.

Artificial intelligence (AI) will also begin to come into focus, to make creating content much easier, and in less time.

Christian Karasiewicz is the Content Marketing Manager at StreamYard and host of the Launch Your Live show, where he shares streaming tips and gear recommendations. He is the host of Social Chatter, the longest-running social media marketing talk show from Social Chefs.

84. Erin Cell

We will see a change in content creation as AI becomes more advanced.

We are already seeing AI images, AI copywriting, and AI voices being created.

I think we will see people combining all of the AI tools to create an entirely new world of content creation.

We will have a difficult time differentiating what is real and what is created by AI.

Erin Cell has been in the blockchain and cryptocurrency industry for five years as the Producer for the Bad Crypto Podcast and The Nifty Show. She has spoken internationally on NFTs & crypto including NFT.NYC, NFT.London & NFC Lisbon. Her focus has been on bringing more women in the NFT space. In early 2022 she launched a new YouTube channel and podcast, The NiFTy Chicks, with her co-host JeNFT. The show is focused on shining a light on women in the NFT space.

85. Jeff Howell

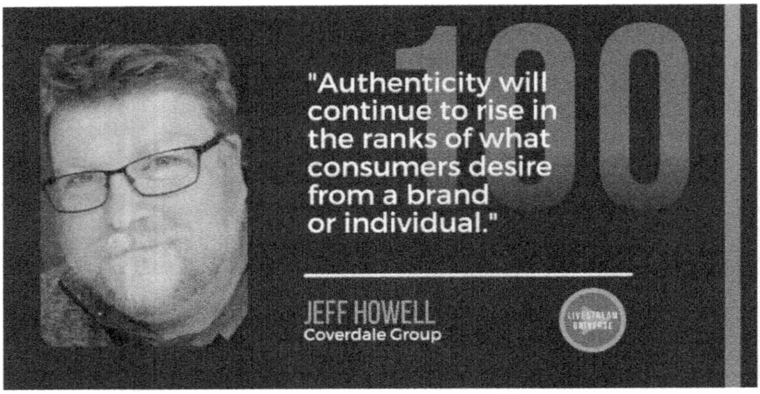

My first prediction is that authenticity will continue to rise in the ranks of what consumers desire.

Whether from a brand or an individual, people will no longer tolerate forced or insincere messaging.

My second prediction, from an e-commerce perspective, is the brands, businesses or individuals that create the fewest barriers for users to pay for goods and services will gain much more market share in 2023.

I think people desire convenience when paying for something. The tougher the check-out experience is, the less likely a buyer will return to purchase, or they may abandon their carts.

Third, I sincerely hope that brands expand their idea of what e-commerce is.

There is an opportunity for brands to integrate the e-commerce experience into their consumer relationship, which I think is a weak point at the moment.

Jeff Howell has transitioned his marketing experience and predisposition to authenticity into a successful e-commerce career. His career mission these days is to evangelize the necessity of having a well-executed e-commerce solution integrated into existing marketing strategies while equipping his clients with the knowledge to be successful.

86. TINA K. VALANT

The global livestreaming market continues its unprecedented, exponential growth.

In 2022, the global video streaming market size was valued at USD $473.39 billion and estimated to reach $1690.35 billion by 2029.

Forward-thinking brands, celebrities, retail chains and sports franchises continue to merge onto the livestream highway.

Wise venues and event planners incorporate virtual-attendance options, behind the scenes peeks and/or special contests/offers/promotions to those who engage. They embrace the immense value in accomplishing far more, with less, by livestreaming.

Anyone can reach out and touch prospective customers/friends/followers/subscribers by creatively utilizing social media, livestreaming platforms and/or influencer collaborations.

This mega-trend is here to stay. Carpe diem!

Tina K. VaLant is a professional photographer/image consultant, original creator of award-winning community "Ladies Go Live," Nature lover, dog rescuer and butterfly enthusiast.

87. Jesse Guthrie

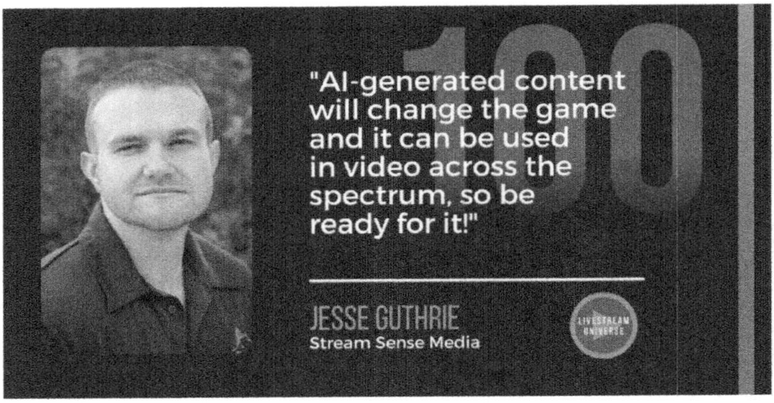

The introduction of AI content into the mainstream! This is going to change the game when it comes to creating content with AI generators (specifically the written text form) that can be used in video across the spectrum.

Scripting for videos — answers to questions for videos — the sky is really the limit here. Be ready for it!

Jesse Guthrie is the owner and operator of Stream Sense Media who helps businesses take their knowledge and get it started on video! Stream Sense Media helps you get in front of the camera and get content out.

88. Tim Gillette

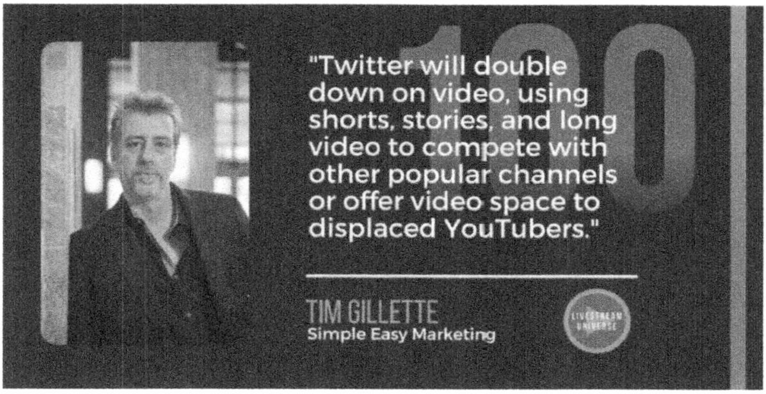

With the stir up at Twitter, I predict in 2023 Twitter will double down on video.

Using shorts, stories and long video. I believe they will do this to compete more with SnapChat, Instagram, and TikTok.

But I also believe they will do this to give users a place to put videos for people who cannot use YouTube anymore as their broadcast channel.

Tim Gillette is a digital content creator, mentor and author. He is the creator of the Simple Easy Marketing System for entrepreneurs to get started and get results.

89. Rose Horowitz

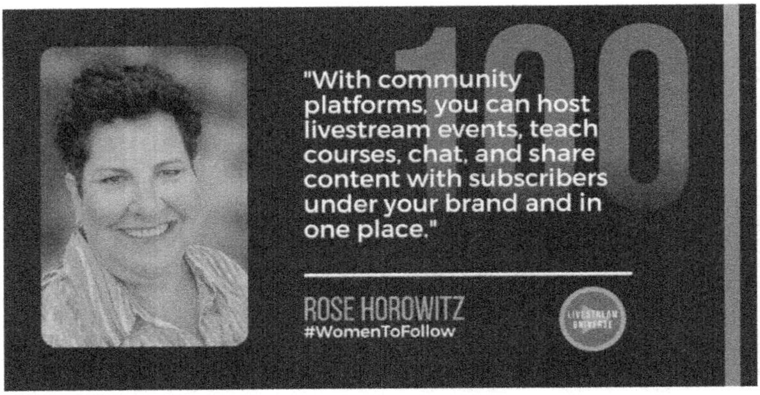

With Twitter in flux under Elon Musk's ownership, shifts on other platforms and advances in AI driving change in marketing, we will see the growth of alternative platforms.

We will see a good deal of change and, in fact, disruption in digital media in 2023.

Creators have been jolted by uncertainty in the digital space.

If an important platform such as Twitter should change drastically or in the worst-case scenario disappear, what happens to the hundreds or thousands of followers a writer or creator has culled and built over the years? How do they retain their community?

Even if you have a big brand, you don't own your community on YouTube, TikTok, Twitter, Facebook, Instagram, or LinkedIn.

What happens if your account is banned for some reason?

With community platforms such as Circle and Mighty Networks, you can host livestream events, teach courses, chat and share content with your subscribers under your brand in one place.

Being an early adapter is key to succeeding in livestreaming and digital media.

Keep on top of trends and check out the new things people are trying. Weigh the positives and negatives.

Digital media will change in significant ways this year so be informed and ready to experiment.

Rose Horowitz is a Pulitzer-nominated journalist and founder of #WomenToFollow, a global hashtag trend that reached 12 million views in its first two weeks. Rose is a thought leader on women's issues, has created content for major corporations, and hosts the #WomenToFollow livestream show.

90. Armel Beaudry Kembe Tchemeube

We will see existing social media platforms experiment more with audio. There will be new audio creator tools extensively making use of AI to lower the barrier of entry for content creators in the audio space.

Those new tools will drive more experimentation in the podcasting space and will drive the emergence of new audio formats.

The debate around what should be considered a podcast and what should not be considered a podcast will grow

Armel Beaudry Kembe Tchemeube is currently the CEO and founder of Trebble FM, an online tool that makes it easy to create and edit audio content.

91. Daniella Malfitano

My prediction is that 2023 will be the year that media marketing will begin to see a mass exodus of participants and engagement as more and more people are getting off social media and are seeking more one on one connections.

Daniella Malfitano is a chef, producer, host, author, activist and speaker helping the world get healthier one tasty conversation at a time!

92. Marco Novo

Curiosity is key if you want to succeed. Don't get me wrong, I'm not talking about getting rich or famous. This is about improving some skill, getting good grades at school, increasing your business results and so on.

And where do livestreams fit into the equation, you're probably wondering? Since I started going live back into 2015, I've learned a lot, always led by my curiosity.

Yes, as a content creator, learning-curiosity is a partnership that goes hand-in-hand all the time, but when we talk about livestream, the music plays loud and energetic.

You need to improve your technical and personal knowledge. You want to shine in front of your audience so you need to prepare yourself. Knowing the gear and the tech is so important and will bring to you new and unexpected knowledge. You will love to see what you'll discover.

Think about everything you can learn from guests and co-hosts during the journey. A friend of mine says all the time, with the guests I already had—more than 200—you earned a master's degree.

Learn Marketing skills with the goals of knowing how to promote your content, how to make it more impactful, how to get the most from it and how to adjust it to your business.

You will learn from your audience, which hopefully will become your followers, customers and friends, depending on your goals. With that, you can adapt, improve and get to know them better.

Emotional/social skills are what you need to deal with so many things while you're live. Please don't get scared as you can do this. Follow the conversation, give feedback and acknowledge interactions. Yes, livestreaming is about multi-tasking, but in a good way.

And don't forget, it's important to keep your seat in the audience. An amazing way to learn is from the speakers, what they are discussing, how they do things and the questions/comments from other viewers.

Livestreaming is an "out of the comfort zone" format or content type, but at the same time the one that open more horizons.

You don't believe me? Well, I was born in Portugal. I natively speak Portuguese. Can you guess how I mostly improved my English? I live in a small village with around one thousand people, far away from the big Portuguese cities. How do you think I gained an invitation to be part of this project with all these amazing marketing, social media, and livestreaming specialists from all over the world?

You got it — with livestreaming, and yes, of course, my curiosity and will to learn.

Are you in?

Marco Novo is an Amazon Influencer and Livestream Producer and Trainer.

93. Marisa Cali

Short, easily consumable stories and video content snippets will be prioritized across various digital channels especially LinkedIn.

Creators can utilize long-form video to be innovative and edit their content into smaller pieces of consumable content.

People will look for content to consume and creators will be expected to share their expertise over and over again.

Marisa Cali has managed and grown her own consulting business since 2015 and embraces the benefits (and challenges!) of the work-from-anywhere lifestyle. In April 2020, she launched the Live on Social Now Productions division of her consulting business, which blends her love of tech, project management, and creativity.

94. Omar Colom

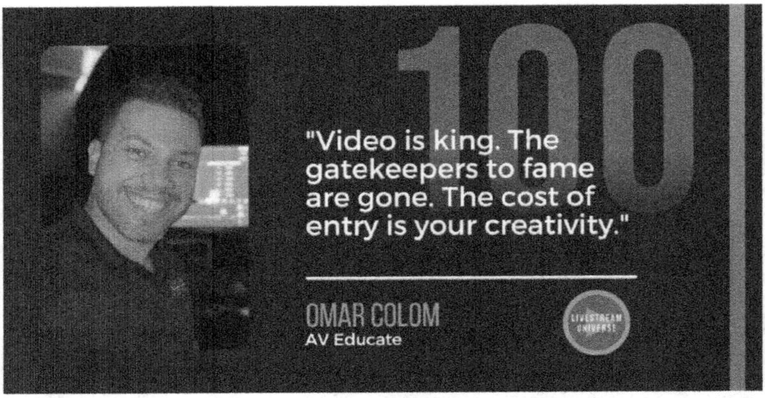

Video is king. The gatekeepers to fame are gone. The cost of entry is your creativity. You are now judged by popular opinion, and that's the good news.

In recent years there has been an increase in livestreaming for various purposes. The type of content that is most successful for livestreamers can vary depending on the audience and the specific goals of the streamer.

However, some general categories of content that have proven to be popular among livestreamers include:

Gaming: Livestreams of video games, including both casual gameplay and professional tournaments, are a popular genre of content.

Entertainment: Livestreams featuring music, comedy or other forms of entertainment can be popular with audiences looking for a more casual and relaxed viewing experience. (Also, a great way to get watching hours.)

Education: Livestreams focused on teaching a particular skill or subject can be popular with viewers looking to learn something new. (This is the category I fall into.)

Professional Events: Livestreams of conferences, seminars and other professional events can be popular with people looking to stay up to date on the latest developments in their industry. Some bring in high-end speakers remotely to an in-person event for a different budget or to test the waters for future events before going to an in-person event. (This is part of what I do for a living.)

Thanks to many factors from 2020, this trend has been driven partly by the widespread availability of high-speed internet and the proliferation of devices such as smartphones and tablets that are used to access or create streaming content. (More on the phone in a minute.)

Another item to be aware of is the increasing use of interactive features in livestreaming, such as the ability for viewers to ask questions or make comments in real time. This interaction creates a more engaging and immersive experience for viewers and helps to foster a sense of community around the stream; if you are reading this book, then you already know the fact what a livestreamer can use to stand out in the crowd and work with the algorithm.

Quality Content: A livestreamer who consistently produces high-quality content that is interesting, informative, or entertaining is more likely to attract and retain a large audience. Neglecting technical considerations such as ensuring that their stream is stable and high-quality or that the audio sounds warm and loud enough to the viewer.

Failing to pay attention to the technical aspects of quality can lead to a poor viewing experience.

Personality: A livestreamer with a unique personality or perspective can differentiate themselves from other streamers and make their content more engaging. Responding to comments and questions can create a sense of engagement and make viewers feel like they are part of the stream.

Interactivity: Livestreaming platforms often allow viewers to interact with the streamer in real-time through features such as chat or Q&A. Streamers who can effectively engage with their audience and create a sense of community around their stream are more likely to stand out. (AV Tech Talks under AV Educate worked these three angles for almost two years and grew in an industry that thrives on being behind the scenes.)

Professionalism: Livestreamers who are well-prepared, organized and able to effectively manage their stream are more likely to attract and retain a large audience.

Marketing and promotion: Livestreamers who can effectively promote their stream through social media, online communities or other channels are more likely to attract new viewers. This comes down to a simple formula: create long-form content, then chop it all up into bite-size content and share it everywhere to bring people back to the long-form content or your page.

Moving forward in the industry, we have to first go back a moment for perspective and to understand where I see the streaming industry going.

Since my niche is education, my prediction will focus on how this applies to educational streamers. The concept of augmented reality (AR) dates back to the 1960s.

But it was in the 21st century that AR technology began to be developed and refined in a way that made it more widely available and practical for everyday use.

Here is a rough timeline of key events and milestones in the rise of AR:

In the 1960s, computer scientist Ivan Sutherland coined the term "augmented reality" in a paper describing a head-mounted display that superimposes computer-generated graphics over the user's view of the real world.

The 1990s: AR technology began to develop for military training and other specialized applications.

1997: The term "augmented reality" is used in a New York Times article for the first time.

1998: Thomas Caudell and David Mizell coin the term "augmented reality" to describe a technology that superimposes computer-generated graphics onto the user's view of the real world.

2000: AR technology is first demonstrated on a smartphone, using a Nokia 9110i device.

2010: AR games and apps become widely available for smartphones, thanks to the release of AR development platforms such as ARKit and ARCore.

2016: The Pokémon Go game becomes a huge hit, popularizing AR and bringing it into the mainstream consciousness.

2017: Google releases Glass Enterprise Edition, an AR headset designed for use in industrial settings.

2020: Apple Continues to grow its iPhone and iPads with LiDAR scanners.

The development and use of AR technology have become increasingly accessible, making it easier for individuals and businesses to create and incorporate AR experiences into their products and services.

One factor contributing to the growing ease of AR development is the proliferation of AR development platforms and tools. These platforms provide pre-built frameworks and libraries, and tools for designing, testing and deploying AR applications.

Some popular AR development platforms include ARKit (for iOS devices) and ARCore (for Android devices), and Spark (from Facebook). These platforms have allowed developers with little or no AR experience to create basic AR experiences quickly and easily. Most of all the filters you already use on Instagram, Facebook, Snapchat and many others are based on this technology.

In addition to development platforms, there are also a growing number of AR-specific software tools and frameworks, such as Vuforia and AR Foundation, allowing developers to create more advanced AR experiences with greater functionality and customization.

Another factor that has made it easier to do AR is the widespread adoption of AR-enabled devices, such as smartphones and tablets. These devices come with built-in sensors and cameras that enable them to track the user's location and recognize objects in the environment, making it possible to create AR experiences that anchor to the real world.

Overall, the increasing availability of AR development platforms, tools and devices has made it easier for developers and businesses to create and incorporate AR experiences into their products and services. It has paved the way for the wider adoption of AR technology.

AR is limitless in the mind of the imagination and in this writer's mind. It's the next digital frontier to bring streamers closer to their viewers.

My vision is to take physical objects, make them into digital objects and ease that barrier to getting hands-on experience when learning.

Omar Colom is the founder of AV Educate and The National Director of Education of Evolve Technology. Omar holds over a dozen certifications and train the trainer certifications in various disciplines in AV. Making him one of the top resources in education for Live and Corporate Events. a decorated combat veteran of the U.S Army and second-generation video engineer who is passionate about helping others achieve their educational goals. A proud father, he is dedicated to providing quality education and resources that enable his students to thrive in their careers. He has a wealth of experience in the audio/visual industry and is known for his dedication to his students' success. As a result, he has helped countless individuals achieve their dreams of working in the audio/visual field. Whether you are just starting out or are looking to take your career to the next level, Omar Colom is an excellent resource for anyone interested in pursuing a career in audio/visual technology.

95. Claudia Santiago

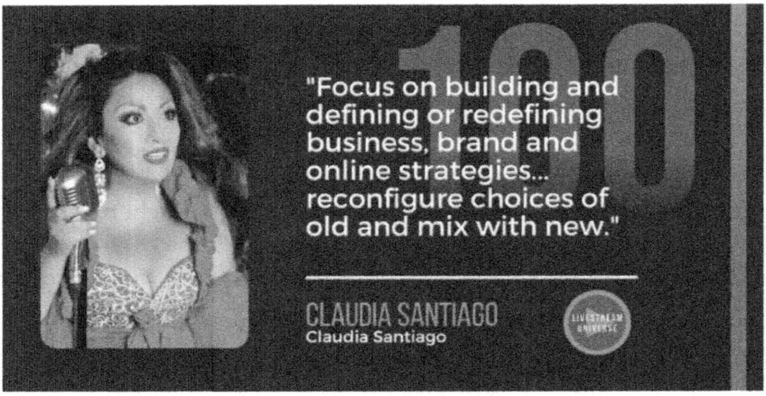

I'm going to focus my prediction this year on building and getting back to the drawing table sorta speak to mean defining or redefining business, brand and online strategies.

Through COVID, businesses that were not online realized it was time to invest in this — great news as we see businesses flourishing both offline and online!

It may be time for some businesses to reassess their strategies, and even the platforms they once used, and refocus on the new (or reconfigure) choices of old and mix with the new.

Platforms, systems and tech may come and go, but there's one thing to always keep at the forefront. Being and staying in a community is still one of the foundations.

And I encourage all to not isolate but continue forming strategic partnerships and walk with those that have been long-term supporters. One of my business mentors told me years ago, human resource is one of the most valuable things we have. Invest in people wisely!

Some businesses may benefit from branching out into hybrid architecture strategies while others may want to scale back their product and service offerings.

It's time to reassess and prepare well.

I see that we may yet face incredible changes in the coming year(s) and it would be of benefit to make strategic planning decisions and carefully prepare for drastic changes in the economy.

This is not heavy news, but good news for any entrepreneur as it's opportunity to turn on a dime in a wise and prepared way!

Like we say in the entertainment industry - Let's get out there, take your stage, and rock it friends!

Claudia Santiago is a Chilean born, Canadian international recording artist, speaker, producer, and entrepreneur. Claudia can be found gracing various stages at her shows with her musicians and online through her livestream broadcasts! She is the founder of Viva-La-Vida! Success - Equipping, Empowering and Educating leaders for career and life success.

96. Jeffrey Bradbury

In 2023, we will see the rise of not just micro-blogging, but the overabundance of micro-video platforms.

Sites like Tik-Tok and YouTube will continue to lead the way, but as time goes on other copycat sites will rise up to saturate the 60-second video market.

Jeffrey Bradbury is an educator and podcaster and the creator of the TeacherCast Educational Network.

97. John Piper

I do believe there will be amazing growth in niche learning opportunities utilizing digital courses, live events and podcast style audio to complement the learning process.

Technology has enabled those who have experience and can share their story in a compelling way that the student can relate to will win and grow others.

I'm personally working on growing my own email list to not need to rely on social media platforms to communicate to my target audiences.

John Piper started his podcasting, livestreaming and media business in 2017. He's passionate about helping others get confident in telling their stories online. He's been a technology leader for over 2 decades based in Dallas, TX.

98. ANITA SONYA

Artificial Intelligence (AI) brings livestreaming and podcasting to a new level of experience for both users and viewers. I can say that AI transforms the livestreaming and podcasting industry. It means that the livestreaming and podcasting platform providers and destinations also need to accelerate this process.

AI, ML (Machine Learning) and DL (Deep Learning) can't be separated anymore from your daily tasks online for media productions as a digital creator.

In many ways, AI will improve the quality of your livestreaming and podcasting, especially for your video and sound. It will help you, from brainstorming ideas to evaluating your content after being published online.

Despite the debates and worries about authenticity among online content creators, we can't deny the fact that AI makes our work easier in creating your content, recorded, or livestreamed, for your business or other purposes. It does not matter what your background is or what industry you are in.

Of course, there is always controversy when a new thing comes up and becomes a big influence globally, especially when strong Artificial General Intelligence (AGI) starts to take over our life since it can make decisions and solve problems like the human brain without our interference.

For now, look at the positive side that it will bring a lot of advantages for livestreamers and podcasters. Not only in digital marketing but also in any field whether that is health, education, personal safety, entrepreneurship, etc.

Anita Sonya is a livestreaming coach and consultant helping business owners transform their businesses through livestreaming. She organizes multiple international summits, including the Online World Summit for Livestreamers and 100 Days with Sonya, bringing together top speakers from diverse industries. Her motto, "Inspire to Succeed," has given her the chance to create platforms supported by international sponsors for the experts in their fields to share their expertise, experiences, knowledge, skills, tips, and tricks to reach the top.

99. MEIKO S. PATTON

Decentralized Social (DeSo) media will dominate in 2023.

Content Creators will not only demand ownership of their data but reserve the right to get paid for what they post without social media companies getting rich off their back.

Marginalized content creators are at the mercy of biased algorithms on social platforms. DeSo completely changes that.

Every post you share on DeSo has the potential to make creators real money. Simply share and earn.

Meiko S. Patton is the Creator of The Daily DeSo. She helps creators capitalize and earn money on decentralized social media.

100. Larry Roberts

Podcasting saw a tremendous rise in consumption in 2022. Along with that increased consumption came massive diversification, welcoming new audiences — narrow or broad — from across the globe.

In 2023, not only will we see that trend continue, but I believe we're going to see it explode.

With this diversification, new opportunities to connect with global audiences will be introduced, allowing podcasters to engage with previously difficult-to-reach market segments.

For podcasts representing brands, businesses and organizations, we'll see an increased ability to connect with consumers on a personal level, lending a voice to their business and establishing a human connection that will undoubtably lead to extended customer retention and overall profitability.

Larry Roberts has become one of the most sought-after podcast consultants and content creators in the industry. Recently named to Podcast Magazine's 40 over 40 in Podcasting, Larry is also the editor-in-chief of one of the largest podcast industry newsletters, The Podfest Messenger.

101. STEPHANIE GARCIA

In 2023, personalized videos using AI will be a game changer.

Companies will be able to call out their customers by name without any manual recording.

We're talking automatic personalized video messages delivered over personalized emails and branded video pages that really drive engagement and customer loyalty.

Video is already the most effective marketing and engagement content, so adding personalization to that mix is a surefire way to increase customer lifetime value.

Stephanie Garcia is a live video strategist for brands and entrepreneurs–from moonwalkers to MasterChefs–who want to go from unknown to unforgettable. A digital marketing expert with 15 years of ad agency experience, she led paid search advertising for one of the largest search agencies in the United States, served as the Director of Social Strategy for San Diego's leading ad agency, and has created powerful campaigns for Fortune 500 companies and small businesses. Today, Stephanie has helped thousands of her clients and students get on the Fascination Fast Track™ to ignite their ideas and become confident on camera.

102. Nancy Ancowitz

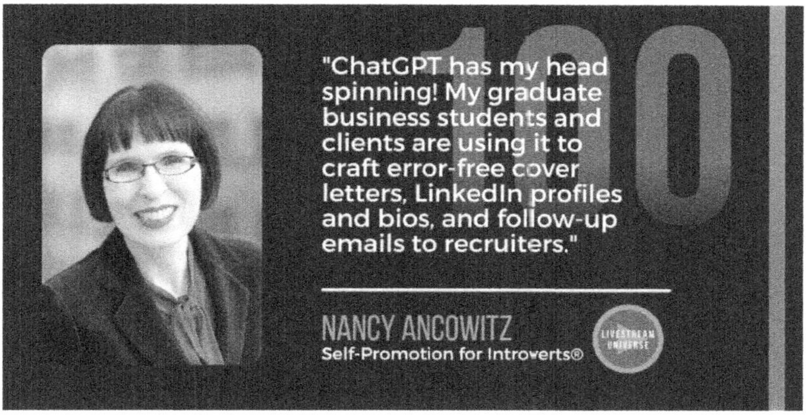

Online chatbot ChatGPT has my head spinning! It's a dizzying gamechanger in multiple ways.

My graduate business students and clients — the majority of whom are not native English speakers — are buoyant as they query it to craft error-free cover letters, LinkedIn profiles and bios, and follow-up emails to recruiters. I love giddily discussing with them how they will use ChatGPT in their job searches and workplaces.

Oh, it can do it in multiple languages, too. On the spot. Make it tighter. Have it rhyme! Tell a joke! I digress from the use case here: career readiness.

I'm wondering, reeling, curious about ChatGPT — and recognize the controversy surrounding it. Still, it's here to stay. So, I'm stoked to think what we'll do with it next.

Nancy Ancowitz is a coach specializing in career advancement and presentation skills, a career director and adjunct instructor at NYU and the author of "Self-Promotion for Introverts®: The Quiet Guide to Getting Ahead." She blogs for Psychology Today and used to blog for Salary.com.

103. Louise McDonnell

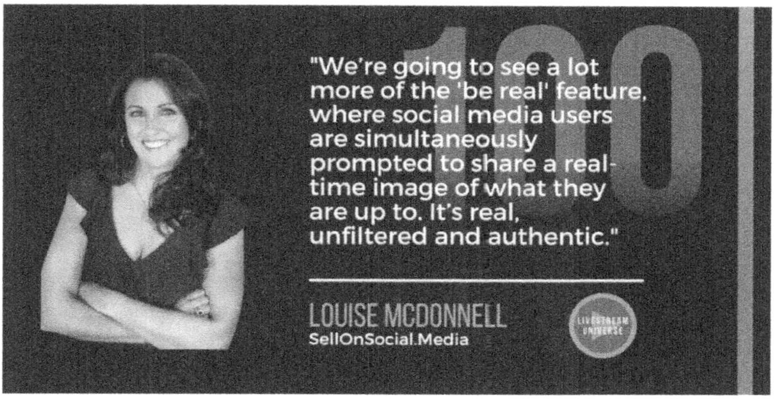

I think we're going to see a lot more of the "be real" feature, where social media users are simultaneously prompted to share a real time image of what they are up to. It's real, unfiltered and authentic.

Launched by the French App, "BeReal," already the feature is spreading to other social media platforms like TikTok, with the introduction of TikTok now. We can expect a continued push of reels and TikTok, and I also think TikTok Ads will become more popular this year.

An award-winning social media trainer and strategist, Louise McDonnell has worked in the digital space since the early 1990s. She is the founder of SellOnSocial.Media, a digital marketing agency operating under two divisions. The Agency services are for businesses and non-profits that wish to outsource their social media marketing to the experts. The Online Training Academy trains entrepreneurs in how to optimize social media to achieve commercial stand-out and boost sales.

2023 Edition

PART III
RESOURCES

NEXT STEPS

Thank you for choosing this book.

Stay on top of the latest trends and tactics for creating content that serves your audience and builds your business. It's free and easy to sign-up for the following Livestream Universe offerings:

1. **Subscribe to the #StreamLeader Report**: Our free monthly report delivers livestreaming and digital media info and strategies to keep you ahead of the competition.

   livestreamuniverse.com/streamleader

2. **Participate in the Monetize Livestreaming Now™ Facebook Group:** Share your expertise and get answers to your questions about livestreaming, digital media and business in our free Facebook group.

   facebook.com/groups/monetizelive

WHAT'S IN ROSS' CREATOR TOOLKIT?

LIVESTREAMING & VIDEO PRODUCTION SOFTWARE

Cloud-Based Livestreaming Software

StreamYard: Exceptional stability unmatched by any other browser-based livestreaming studio. Improved quality for editing and podcasting thanks to the addition of local recording with separate audio and video tracks for each participant. New "StreamYard On-Air" webinar platform is a winner.

livestreamuniverse.com/streamyard

Livestreaming Software for Mac

Ecamm Live: Release your inner creative and build your scenes out the way you always wanted. Lots of user control without the learning curve required by Wirecast or OBS. The team provides helpful content for learning different ways to use Ecamm Live.

livestreamuniverse.com/ecamm

Mobile Livestreaming Software for iOS

Switcher Studio: Remains the industry standard for iOS live video production. Use your iPhones and iPads for multi-cam video, insert titles and graphics, edit live and stream anywhere. New features to support live shopping.

livestreamuniverse.com/switcherstudio

Professional Livestreaming Software

Wirecast: If you want to produce a TV-quality livestream without breaking the bank, Wirecast is a robust, professional livestreaming solution that gives you complete control.

livestreamuniverse.com/wirecast

Video Editing & Screencasting

ScreenFlow: While I continue to experiment with text-based editing, ScreenFlow remains my trusted tool for video editing and screencasting. More advanced than iMovie and easier to use than Premiere Pro or Final Cut Pro.

livestreamuniverse.com/screenflow

Descript: This text-based video and audio editing software is a revolutionary advancement that both simplifies the editing process and contains powerful features for upgrading the quality of your audio and presentation of your video. It transcribes your video or podcast automatically upon upload. Whether as your sole editing tool or in combination with a traditional video editor, this is a game changer.

livestreamUniverse.com/descript

YouTube Optimization

TubeBuddy: If you don't have the free Chrome browser extension, follow the link below and download it now. Paid plans give you the full benefit of this essential tool for growing your YouTube channel, researching topics and getting more eyeballs to your videos.

livestreamuniverse.com/tubebuddy

Video Monetization

Uscreen: It's time to monetize livestreaming and on-demand video. Uscreen makes it easy to sell subscriptions, courses and membership communities. Run your own Netflix-style streaming service.

livestreamuniverse.com/uscreen

Video Overlays & Backgrounds

Streamer Templates: Stand out on live video with customizable templates designed specifically to work with StreamYard's layouts.

livestreamuniverse.com/templates

PODCAST SOFTWARE

Podcast Media Host

Captivate.fm: Designed with the podcaster in mind, it's an unmatched user experience with all the tools you need to manage your podcast or network. It's my favorite podcast hosting service.

livestreamuniverse.com/captivate

Podcast Recording & Remote Interviews

Squadcast: Record audio & video interviews in studio quality from anywhere, with anyone, around the world. Create engaging audio and video content with an intuitive platform that allows remote recording and simplified editing.

livestreamUniverse.com/squadcast

Podcast Website

Podpage: Build a beautiful podcast website in five minutes from your RSS feed. Lots of customization options to build a multi-purpose fully functional website like the one I have at RossBrandRecordings.com. Unbelievably affordable premium plans include hosting and custom domains.

Livestreamuniverse.com/podpage

AI TOOLS

Transcripts & Audio/Video Editing

Otter.ai: AI transcripts of audio and video content.
livestreamuniverse.com/otter

Descript: AI transcripts and audio/video editing. See "Video Editing & Screencasting" above for more.
livestreamUniverse.com/descript

Copywriting

Jasper: Use the help of AI to write high converting copy for increased conversions and higher ROI. Templates for everything from blogs to video scripts.

Livestreamuniverse.com/jasper

AUDIO & VIDEO GEAR

Links to video gear below at

kit.co/rossbrand/video-gear-recommendations.

Studio Microphones (XLR)

Shure SM7B: While it rocketed to popularity in recent years among podcasters and streamers, it's been a radio and music industry staple for decades. It's the mic I use in my home studio. I love its versatility and warmth.

Electro-voice RE20: Though not as popular in the digital media space as the SM7B, this was my favorite mic when working in radio. It deserves far more attention than it gets from podcasters and live video hosts. The new black edition looks good on video too.

Heil PR40: Not the choice for every voice, but on some voices this professional broadcast mic sounds amazing. Not nearly as gain hungry or as expensive as the mics listed above.

USB Microphones

Samson Q2U: If you're new to livestreaming and podcasting and don't have mixers and audio interfaces sitting around your home, then there is no reason to start with any other mic. It sounds far too good for the price and nobody I've recommended it to has ever complained.

Shure MV7: If you want the option of plugging in USB but need to sound like you're using one of the better studio mics, then this hybrid is an excellent choice. Not sure there is a better sounding dynamic USB option.

Elgato Wave 3: Some people want to keep the mic out of their shot or at least not as close as most of the options above require. Although it goes against my principles to recommend a condenser mic for home studio work, this is a strong choice.

Remote Interviews & Field Reporting (XLR)

Electro-voice RE50: When I need to go live from events or capture interviews in noisy environments, this is my top choice and I really don't have a runner up. There's a good reason — it's the broadcast industry standard microphone for electronic newsgathering.

All-Purpose Microphone (XLR)

Sennheiser e835: Want one affordable, professional mic that can be used in the studio, out in the field and on stage? This remarkable handheld dynamic mic travels with me whenever I'm on the road, usually right next to the RE50.

Audio Interfaces, Mixers and Recorders

Sound Devices MixPre-3: It's got the best sounding preamps of any device in its class and its analog-to-digital converters are excellent as well. The MixPre makes a wonderful audio hub in your home studio and an exceptional field recorder for video shoots. There's a learning curve to using it, but the reward is professional-quality audio.

Rode Rodecaster Pro 2: This soundboard-like device just looks like a lot of fun to use. It's become the go-to choice for many podcasting studios. If you are producing while livestreaming or recording, this mixer/interface/recorder is a good blend of quality and ease of use.

Zoom Podtrak P4: Not as cool as the Rodecaster Pro, but light-weight and portable with high-quality audio at a fraction of the price.

Focusrite Vocaster: Both new and experienced podcasters and livestreamers will love the ease of use and smart feature set included in the Vocaster interfaces from Focusrite. More than enough gain to power an SM7B or RE20, it also has auto gain, vocal presets and can send audio into your camera.

Headphones

Sony MDR-7506: I use these every day. They're comfortable, work great for editing and are one of the better aesthetic choices if you wear headphones on video.

Microphone Boom Arms

Rode PSA-1Boom Arm: Perfect for heavier studio mics like the SM7B, RE20 and PR40.

Neewer Mic Boom Arm: Works well with handheld mics such as the Q2U and e835 at a much lower price than the Rode.

Webcams

Logitech c922x (or c920): These webcams are all you need to go live or record with good quality video.

Mirrorless Camera

Sony A6400: When it comes to mirrorless video cameras, especially for livestreaming, Sony is a winner. Many worthy options from Sony, but I went with this one and couldn't be happier with it.

Green Screen

Elgato Green Screen: "What green screen do you use?" That's one of my most frequently asked questions over the past few years. I don't have enough data to compare it to others on the market, but I can say that it came highly recommended by photographer and video creator Mike Murphy (contributor to Volume 2), it's well-made and I have no complaints. It's also easy to setup and take down in your home studio.

GRAPHICS & SHORT VIDEOS

Streamer Templates: Stand out on live video with customizable templates designed specifically to work with StreamYard's layouts.
livestreamuniverse.com/templates

Canva: Popular DIY graphic design platform that does so much more than provide templates for static social posts. Edit short-form videos, design your website, schedule your social posts and write out articles with AI. The background removal tool is worth the price of the Pro Plan by itself.
livestreamuniverse.com/canva

PlaceIt: Create mockups, logos, social posts and videos in seconds. Great for intro videos for livestreams.
livestreamuniverse.com/placeit-2

Wave.video: Create thumbnails, backgrounds, overlays, intros, outros and other image and video assets to use in your content.
livestreamuniverse.com/wave

DepositPhotos: High-quality stock photos and video b-roll.
livestreamuniverse.com/depositphotos

Creative Fabrica: Fonts and design elements for your online graphics and physical products.
livestreamuniverse.com/creative-fabrica

WEBSITE, EMAIL & MARKETING

WordPress Hosting

Rocket.Net: A last minute addition to the book, Kim Doyal recommended Rocket.Net when I was searching for a new webhost because I couldn't face publishing another book with an embarrassingly slow website. I'm absolutely blown away by how blazing fast it is since changing hosts. Rocket.Net completed the move for me at no extra charge within an hour from when I signed up.

livestreamuniverse.com/rocketnet

WordPress Themes & Page Builders

Elementor Builder: Build and customize every part of your website intuitively with Elementor's visual builder. Zero coding required.

livestreamuniverse.com/elementor

Elegant Themes: Divi is the most popular premium WordPress theme in the world and comes with Divi Builder.

Livestreamuniverse.com/divi

Podcast Website

Podpage: Build a beautiful podcast website in five minutes from your RSS feed. Lots of customization options to build a multi-purpose fully functional website like the one I have at RossBrandRecordings.com. Unbelievably affordable plans include hosting and custom domains.

livestreamuniverse.com/podpage

Domains

NameCheap: Tends to have lower prices than its competitors along with a beautiful website, convenient mobile app and easy-to-use interface.

livestreamuniverse.com/namecheap

Email Marketing

ActiveCampaign: Email list building on the world's leading email automation platform.

livestreamuniverse.com/active

Course Delivery & Sales

Thinkific: Popular choice for building online courses and membership communities.

livestreamuniverse.com/thinkific

Social Media Management

Agorapulse: The premier tool for social media scheduling, listening, engaging and analytics just got even better with the ability to schedule TikTok videos right from inside Agorapulse. Great for teams and collaboration. Indispensable when managing many accounts. Customer service is top-notch.

livestreamuniverse.com/agorapulse

Content Planner

Create It: A planner that helps you develop a content strategy that supports your business and your life.

Livestreamuniverse.com/createit

ACKNOWLEDGEMENT

While the *100 Predictions* series is a highly collaborative effort thanks to the contributions of 100+ top digital creators, industry insiders and online entrepreneurs, the editing and graphics have been mostly a solo endeavor going back eight years to the first Livestream Universe Predictions blog post. What makes this volume of predictions different is the addition of an editor working with me. I brought Sue-Ann Bubacz of Write Mix for Business, who provided valuable suggestions regarding my intro and epilogue for Volume 2, onboard to help with editing. It's been a joy to work with Sue-Ann, whom I met in 2020 – you guessed it – during a livestream. She's not only meticulous and conscientious with a great work ethic, but she quickly picked up on my editing style for the Predictions. She also upgraded the look and feel of the graphics. Thanks, Sue-Ann, for making this a fun and successful project.

A huge thank you as well to Chris Stone and Jim Fuhs of Dealcasters Live for writing the book's foreword. Chris and Jim are remarkably generous in sharing their expertise regarding Amazon Live and the Amazon Influencer Program with anyone interested in diving into live shopping and shoppable videos. They truly understand that as livestreamers and creators, we all do better when we collaborate, as bringing more good hosts into the mix results in more viewers (and commissions) for all of us. They've never stopped telling the world that my video on how to join the Amazon Influencer Program started them on the path to hosting Dealcasters Live. While true, the bigger story is what they did after watching that video – they elevated the content on Amazon Live by combining guest interviews with live selling in an entertaining talk-show format.

Thank you to Dale L. Roberts for continuing to share your self-publishing know-how with the world. I've followed your game plan starting with the first volume and the results speak for themselves: Eighteen book awards and a #1 Amazon Best Seller in six countries. If you are serious about self-publishing, I highly recommend exploring Dale's livestreams, podcasts, YouTube videos, books, courses and consulting. He's the best at what he does!

Thanks to Rachael Anne Clifton aka "Beauty Bubble" for always lending your ears and sharing words of wisdom. You have an amazing ability to take good ideas and pinpoint the missing ingredient that can make them great. You also catch the bad ideas before time and money are wasted.

Thanks to Danielle Camorlinga for your talent and discernment as the designer of the outstanding book covers in this series. It's such a relief knowing I can count on Danielle to have a book cover that grabs attention while being a perfect representation of the *100 Predictions* brand. It's an honor to work with you and I hope you are able to do this for as many more Predictions books as we ultimately publish.

Thanks to those there at the beginning and throughout all eight predictions, including three books: Jennifer Quinn "JennyQ," the guest on my first livestream show in 2015 and author of the foreword to the first volume; Mitch Jackson, who pioneered livestreaming in the legal profession and is now leading us all into Web3; and Jennie Mustafa Julock aka Coach Jennie, who continues to be platform agnostic while enjoying community and growing her business through whichever medium she works in.

Thanks to Mark Babbitt, a contributor to the first Predictions blog post and the first social connection to reach out and meet with me in person a few years prior to livestreaming.

Thanks to Jon Burk and Joie Gharrity – the Hollywood wing of the Livestream Universe, although I'm not sure if you two have ever met – for your enthusiastic support of the Predictions over the years. Thanks to Rob Greenlee and Jessika Phillips for always going above and beyond with your contributions.

Thanks to live video host and podcast producer Caren Glasser for sharing my enthusiasm for testing new digital platforms. Look forward to more collaborations with you.

Thanks to Podfest founder Chris Krimitsos for shining a spotlight on authors at your podcasting and video events.

Thanks to Hannah Jacobson and the team at Book Award Pro for all you do to support authors.

A special thanks to Tim McDonald, Kim Doyal, Dave Jackson, Nancy Ancowitz, Vincent Suppa, Michael Kinney, Jeffrey Fitzgerald, Valerie Morris, Rick Mayer, Eric Hunley, Bee Smith, Jeff Sieh, Chris Curran, Brad Friedman, S. Chris Edmonds, Dr. Tachi, Dan Currier, Ian Anderson Gray, Bridgetti Lim Banda and Kevin Kolbe, Renee Hastings and Mae Esther Maralit.

Thanks to all the contributors to this book for sharing your insights and expertise. Each of their names appears in the table of contents and at the top of their chapter.

And last but not least, thanks to my parents for always being my rock. I love you.

ABOUT THE AUTHOR

Ross Brand is the connection between brands, early-stage startups and customers through the power of live video. He's served as an on-air personality, brand ambassador and consultant for industry brands such as StreamYard, BeLive, LiveU and Wirecast. Ross was instrumental in the rapid growth and adoption of StreamYard through his shows and network of digital influencers. Switcher Studio selected him as one of "5 Live Video Experts to Follow" and Klout named him the world's #1 livestreaming expert on social media.

Ross is the author and founder of the *100 Predictions* franchise. His book series, *100 Livestreaming & Digital Media Predictions*, reached #1 Amazon Best Seller status in six countries and won 18 book awards. He's been featured by Nasdaq TV, Forbes and Psychology Today.

Relevant Links:

Website – livestreamuniverse.com

Podcast – rossbrandrecordings.com

Twitter – twitter.com/irossbrand

Instagram – instagram.com/rossbrand1

LinkedIn – linkedin.com/in/rossbrand

YouTube – youtube.com/livestreamuniverse

Pinterest – pinterest.com/rossbrand

TikTok – tiktok.com/@rossbrandlive

Facebook Page – facebook.com/livestreamuniverse

Facebook Group – facebook.com/groups/monetizelive

Made in the USA
Monee, IL
13 September 2026

62981959R00134